THE OPEN UNIVERSITY

Science:
A Second Level Course

9 and 10   Analysis of Populations

S299
GENETICS

Prepared by a Course Team for the Open University

THE OPEN UNIVERSITY PRESS

Course Team

Chairman and General Editor
Steven Rose

Unit Authors
Norman Cohen (*The Open University*)
Terence Crawford-Sidebotham (*University of York*)*
Denis Gartside (*University of Hull*)
David Jones (*University of Hull*)
Steven Rose (*The Open University*)
Derek Smith (*University of Birmingham*)
Mike Tribe (*University of Sussex*)
Robert Whittle (*University of Sussex*)

*Consultant

Editors
Ian Nuttall (Units 3–5, 9 and 10)
Jacqueline Stewart

Other Members
Bob Cordell (*Staff Tutor*)
Mae-Wan Ho*
Jean Holley (*Technician*)
Stephen Hurry
Roger Jones (*BBC*)
Aileen Llewellyn (*Course Assistant*)
Michael MacDonald-Ross (*IET*)
Jean Nunn (*BBC*)
Pat O'Callaghan (*Evaluation*)
Jim Stevenson (*BBC*)
* From January 1976

The development of this Course was supported by a grant from
the Nuffield Foundation.

The Open University Press,
Walton Hall, Milton Keynes.

First published 1976.   Reprinted 1979.

Copyright © 1976 The Open University.

Designed by the Media Development Group of the Open University.

Set by Composition House Ltd, Salisbury, Wiltshire.

Printed in Great Britain by Eyre and Spottiswoode Limited,
at Grosvenor Press, Portsmouth.

ISBN 0 335 04287 2

This text forms part of an Open University Course. The complete list of Units in the
Course appears at the end of this text.

For general availability of supporting material referred to in this text please write to the
Director of Marketing, The Open University, P.O. Box 81, Walton Hall, Milton Keynes,
MK7 6AT.

Further information on Open University Courses may be obtained from the Admissions
Office, The Open University, P.O. Box 48, Walton Hall, Milton Keynes, MK7 6AB.

9 and 10
Analysis of Populations

S299
GENETICS

Contents

List of scientific terms used in Units 9 and 10

Developed in these Units	Page No.	Developed in these Units	Page No.
ABO blood groups	400	inbred line	396
agglutination	400	incompatibility systems	452
allopatric	441	industrial melanism	426
allozyme	436	mating system	396
aposematic	457	mean fitness	419
apostatic selection	463	meristic characters	398
assortative mating	413	migration	467
balanced polymorphism	433	mimicry	439
Batesian mimicry	439	model	439
biochemical polymorphism	436	modifying genes	429
breeding system	396	monoecy	450
cline	401	Müllerian mimicry	439
continuous variation	398	mutation pressure	416
cryptic	426	neutral alleles	472
differential segment	449	overlapping generations	414
dioecy	447	pairing segment	449
disassortative mating	413	pin	454
discontinuous variation	398	population size	476
doubling dose	467	preadaptation	432
equilibrium	405	protandrous	450
effective population size	476	protogynous	450
fitness	410	random/randomly mating	413
founder principle	475	relative fitness	410
gametophytic incompatibility	453	S alleles	452
genetic drift	414	segregation load	484
genetic load	484	selection	399
genetic polymorphism	432	selective coefficient	410
gynodioecy	450	selectively neutral	417
Hardy–Weinberg law	407	sex ratio	449
hermaphrodite	450	sibmating	431
heteromorphic incompatibility	454	sporophytic incompatibility	453
heterostyly	454	supergene	442
homomorphic incompatibility	454	sympatric	441
		thrum	454
		transient polymorphism	433

Note for students other than those registered with the Open University

Course S299 Genetics has been designed for use in whole or in part by universities, polytechnics and colleges, as well as the Open University itself. Units 9 and 10, and Units 11, 12 and 13 can be detached readily from the rest of the Course, even though they do rely on some of the ideas and information presented in earlier Units. We suggest that teachers or students who are contemplating using these Units should consult S299 G *An Introduction and Guide to the Course*, so that some understanding of the necessary background required can be gained.

Objectives for Units 9 and 10

After studying these Units you should be able to:

1 Define, recognize the best definition of, and place in the correct context, the items in the list of scientific terms on p. 391.

2 Derive the Hardy–Weinberg equilibrium for the frequency of alleles and genotypes and list the known assumptions implicit in the model.
(ITQs 1, 3, 10 and 18; SAQ 1)

3 Explain why mutation pressure alone is insufficient to account for evolutionary change.
(SAQ 2)

4 Derive and use models for the calculation of frequencies of alleles in a population.
(ITQs 1, 2, 4, 7, 18 and 20; SAQs 3 and 11)

5 Detect and interpret departures from the Hardy–Weinberg equilibrium.
(ITQs 2 and 5; SAQ 3)

6 Understand the modifications of the basic Hardy–Weinberg model to allow for selection.
(SAQ 3)

7 Detect limitations in the Hardy–Weinberg model modified to allow for selection.
(SAQ 3)

8 Describe evidence showing evolution in action.
(SAQ 4)

9 Explain why dominance is a property of characters and not of genes.
(ITQ 8; SAQ 5)

10 Estimate the expected frequency of alleles in a population in which there are three alleles with a simple dominance hierarchy.
(ITQs 6 and 7; SAQ 7)

11 Describe examples of the methods used for the study of genetic polymorphism.
(ITQ 9; SAQs 6 and 8)

12 Give examples of genetic polymorphism and of ways by which selection disturbs the equilibrium situations.
(ITQs 5 and 8; SAQs 6 and 8)

13 Explain, in outline, the gametophytic and sporophytic incompatibility systems in flowering plants.
(ITQs 12 and 13; SAQ 9)

14 Describe the role of frequency-dependent selection in the maintenance of genetic polymorphism.
(SAQ 8)

15 Distinguish between discontinuous variation maintained by a balance of selection and mutation on the one hand and by a balance of selective forces on the other.
(SAQ 11)

16 Calculate the effects of changing the mutation rate on the frequency of disadvantageous recessive alleles in a population of diploid organisms.
(ITQs 19 and 20)

17 Give one example of how an understanding of some of the principles of population genetics can help in the study of migration in humans.

18 Demonstrate how small population size leads to non-systematic changes in allele frequency and evaluate the possible role of such changes in the evolutionary process.
(ITQ 21; SAQ 10)

19 Given the appropriate formula, calculate the rate per generation at which allele frequencies may change.

20 Give evidence of having understood why there are problems with the hypothesis of neutral alleles.
(SAQ 11)

Study guide for Units 9 and 10

In Unit 1 we said 'The basic aim of population genetics is to study genetic variation in populations in relation to evolution, by determining the frequency of occurrence of different genotypes in populations of individuals of the same species and by studying the factors affecting these frequencies'. Having acquired the basic concepts and tools of genetics from Units 2–4 of the Course, you are ready to apply them to the quantitative genetics of Units 9–13.

Units 9 and 10 show the relationship and interdependence of the practical aspects of population genetics and their theoretical framework. Problems like the effect of the eradication of malaria on the frequency of sickle-cell haemoglobin in humans, the genetic effects of doubling the mutation rate as a result of environmental pollution, insect resistance to DDT and the relationship between disease and the ABO blood-group system in humans are considered side by side with mimicry and melanism in insects, the sex ratio in *Silene* (campion), chromosome variation in *Drosophila* (fruit flies) and colour and banding patterns on *Cepaea* (snail) shells.

Theoretical models that have been developed in attempts to explain the variation in natural populations are introduced in the text, and you will be expected to work through examples that demonstrate both the predictive power and the limitations of these techniques.

Nearly all the situations we shall consider in these Units are great simplifications. They have to be for several reasons, but mainly because the largest computer cannot handle all the variables for which we ought to allow, and we don't know what all the variables are anyway. We have, therefore, to make assumptions and temper our interpretations accordingly.

As this is a double Unit you should allocate four weeks to its study. It is assumed that at this point in the Course you will have completed the statistics text*, and in particular will be familiar with Sections ST.2, 3, 4, 5, 8, 10 and 12. Remember that a summary of statistical techniques was included as an appendix to the module and you might find it easier to refer to that in the first instance.

The division between Units 9 and 10 is fairly arbitrary and is only made to make it easier for you to plan your workload. If you have difficulty in completing these Units you should be prepared to omit Unit 13 (Ecological and Evolutionary Genetics). You could also just skim through the Sections on Genetic Drift (10.5).

Some of the topics in these Units lend themselves very readily to step-by-step learning, involving questions and answers, for example mimicry in insects and chromosome polymorphism in *Drosophila pseudoobscura*. A study of blood groups on the other hand, requires the absorption of information derived from parallel studies, with a subsequent synthesis of information towards the end. This means that some Sections will require pencil and paper involvement, with you working through the in-text (ITQs) and self-assessment questions (SAQs) as they are presented. If you do not you may not be able to follow the rest of the text. You should immediately check your answers to these questions against those given on p. 491 (ITQs) and p. 496 (SAQs).

Reference to the Objectives will help you decide what you need to know from the Units.

It is appreciated that some of you will find the arithmetic and algebra heavy-going, although there are really *no* mathematical concepts in these Units that are not included in some current 'O' level syllabuses, and certainly none that do not occur in S100 M and S100 MB†.

All the key equations and mathematical expressions in the text have been numbered and flagged, because we shall be returning from time to time to expressions that we have derived earlier. A fold-out sheet at the end of the Units contains all the formulae referred to more than once in the text.

---

* The Open University (1976) S299 STATS *Statistics for Genetics*, The Open University Press. From now on we shall refer to it by its code *STATS*.

† The Open University (1970) S100 M *Mathematics for the Foundation Course in Science*, The Open University Press; The Open University (1971) S100 MB *A Guide to S100 Mathematics —MAFS for Beginners*, The Open University Press.

To indicate how you should study the mathematical Sections we have used a letter code. You will be required to derive all the equations in the Sections labelled A. Those in Sections labelled B will be used more than once in the text, but it is not necessary for you to learn their derivations, and finally, in Sections labelled C, equations are given which will not be used again in the Units.

Some people see the answers to genetic questions intuitively. For them our explanations will seem tedious, but for the majority we have tried to argue explanations using the following approach. To begin with we have to decide what unambiguous conclusions can be drawn from the information available. These conclusions are then used as the basis for explaining, as far as possible, the rest of the problem.

Do not expect definitive answers to any of the problems we discuss. They are all open ended and complicated; that is why they are so interesting.

In addition to the text, the associated broadcast material should also be studied carefully. TV programme 9 (Studies of natural populations of snails) can be viewed and Radio programme 9 (Isoenzymes in Man) heard before Units 9 and 10 are read. TV programme 14 (Molecular Evolution) deals with some of the topics raised in Unit 10. Radio programme 10 revises the Hardy–Weinberg law and shows some of its uses in human genetics.

Introduction to Units 9 and 10

> A satisfactory theory of natural selection must be quantitative. In order to establish the view that natural selection is capable of accounting for the known facts of evolution we must show not only that it can cause a species to change, but that it can cause it to change at a rate which will account for present and past transmutations.

> Haldane, J. B. S. (1924) A mathematical theory of natural and artificial selection, Part 1. *Transactions of the Cambridge Philosophical Society*, **22**, 19–41.

> The intention of this book has not been to show what sorts of things can be brought in to illustrate mathematical ideas, but, on the contrary, to show the sort of way in which mathematical thinking can illuminate our understanding of some very ordinary things in the world about us and stimulate a more imaginative interest in them.

> The last sentence in Frank Land's *The Language of Mathematics*, Murray (1960).

Population and ecological genetics are concerned with some very ordinary things which are going on in the living world about us, and there is no doubt that the language of mathematics has made invaluable and essential contributions to our understanding of the processes involved. Early in Units 9 and 10 we ask questions that can only be answered by resorting to techniques employing algebra and arithmetic. To those of you who have little difficulty in thinking quantitatively, we expect you will find something here to stimulate your curiosity. To those who feel shy of mathematics, we hope we have posed urgent problems in genetics in such a way that all your previous inhibitions will be cast aside and you will plunge in and possibly even enjoy the elegance and insight that elementary algebra can often give to important questions.

9.1 Variation and heredity

9.1.1 Heritable and non-heritable variation

Genetics is a problem-orientated subject, so let us start these Units with a problem.

> QUESTION Among a family of plants uniformly about a metre high (tall) two individuals were observed that were only 0.5 m high (short). Two of the tall plants and both of the short plants were self-fertilized, and 50 progeny from each mating were sown out the following year, when *all* the plants turned out to be a metre tall. What do you conclude about the way in which the character of height is determined in this family?

> ANSWER We expect that you started with the statement in the first sentence and assumed that an allele determining short plants was recessive to an allele determining tall plants.

> There is no real justification for making this assumption, because it is also possible that the two short plants could have arisen as a result of a new dominant mutation. When we use the assumption of a recessive allele for shortness we find that there is no means of obtaining tall plants when short plants are selfed. On the other hand, when we assume that the two short plants result from newly arisen dominant mutant alleles, we are still thwarted by the information in the second sentence, because on selfing we would expect three-quarters of the progeny to be short.

> Alternatively, we can take the information in the second sentence first. We must conclude that, even on selfing, the character for shortness does not appear among the progeny of short plants and we are led to question whether the character of plant height is genetically determined in this family. Without any further information, our conclusion must be, therefore: 'Not all variation is inherited'!

If you had difficulties with this question you are in very good company because the distinction between heritable and non-heritable causes of variation was often not appreciated by some of the leading early geneticists. Even today there is still confusion.

Because genetics deals with genes it is often assumed that the characters being studied have a simple hereditary basis. The first task of the geneticist, however, is to show whether or not the variation in these characters is determined by heritable differences. By appropriate breeding programmes, it is often possible to answer two questions in one experiment: firstly, is the variation inherited and secondly, if the variation is inherited, what is the nature of the inheritance? If we find that the variation is not inherited, or it is heritable in a manner that cannot be explained in basic Mendelian or quantitative terms, we must then decide whether we are looking at direct effects of the environment or whether we have examples of phenocopying (Unit 8, Section 8.6.2) maternal effects (Unit 7, Section 7.6 and Unit 8, Sections 8.5.2 and 8.5.3) or other extrachromosomal influences. The relative importance of the genetic and environmental influences on the determination of a character varies from one character to another, and depends on both the ancestry of the individual being measured and also on the environment in which that individual has developed. But we can say with a degree of finality most unusual in science that no character differences, not even the most striking, are determined entirely by genes on the one hand *or* entirely by the environment on the other, although there are examples of particular differences between individuals which are almost entirely due to environmental effects.

One of the most elegant demonstrations of both heritable and non-heritable variation in the same character was reported early this century by Johannsen. He was studying seed weight in the bean *Phaseolus vulgaris* and began in 1901 with a mixture of 16 000 brown seeds obtained from many different plants of the variety Princess. In general, seeds obtained from plants grown from the 25 heaviest seeds had a greater average seed weight than progenies derived from the 25 lightest seeds. Johannsen was, therefore, able to select for seed weight because the differences between light and heavy seeds of the same variety were predominantly the result of heritable differences (see Fig. 1).

He then inbred 19 plants, obtaining several generations of progeny by self-fertilization, which is the normal system of mating in this species. Keeping the progeny of each plant separate, he selected from each family the heaviest and lightest seeds, and used these to propagate the next generation. In this way he established 19 *inbred lines*. After 6 generations in his line 19, the seed parents produced from the lightest seeds gave progeny with an average weight of 370 mg per seed, and the progeny produced from the heaviest seeds also had an average weight of 370 mg per seed. Clearly at this stage in his breeding programme the differences in weight between the heaviest and lightest seeds produced by one plant could not be due to heritable differences.

inbred line

QUESTION Starting with a heterozygote *Aa*, and assuming a *breeding system* of self-fertilization and equal progeny production by each plant, determine the rate per generation at which the frequency of the heterozygote declines in subsequent generations.

breeding system

[*Hint*: write down the frequencies of the genotypes in the F_2, F_3 and F_4 generations. Refer to Unit 1 if you have forgotten the frequencies of F_2 genotypes.]

ANSWER We start with an individual of *Aa* genotype (equivalent to an F_1) so the frequency of heterozygotes is one.

		Equivalent to				Frequency of heterozygotes
Generation	0	F_1		Aa		$(\frac{1}{2})^0 = 1$
	1	F_2	$\frac{1}{4}AA$	$\frac{1}{2}Aa$	$\frac{1}{4}aa$	$(\frac{1}{2})^1 = \frac{1}{2}$
	2	F_3	$\frac{1}{4}AA + \frac{1}{8}AA$	$\frac{1}{4}Aa$	$\frac{1}{8}aa + \frac{1}{4}aa$	$(\frac{1}{2})^2 = \frac{1}{4}$
	3	F_4	$\frac{1}{4}AA + \frac{1}{8}AA + \frac{1}{16}AA$	$\frac{1}{8}Aa$	$\frac{1}{16}aa + \frac{1}{8}aa + \frac{1}{4}aa$	$(\frac{1}{2})^3 = \frac{1}{8}$

Thus the frequency of heterozygotes is $(\frac{1}{2})^n$, where n is the number of generations of selfing; that is, the frequency of heterozygotes in any generation is one-half of the frequency in the previous generation when the breeding system (*mating system*) is selfing.

mating system

396

16 000 seeds of the bean variety
Princess derived from many different plants

25 smallest seeds selected   25 largest seeds selected

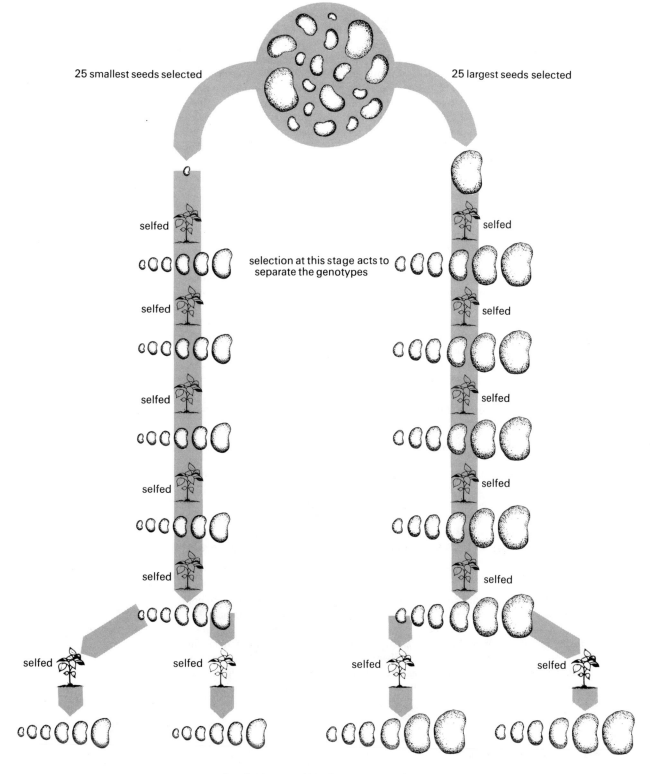

selfed   selfed

selection at this stage acts to
separate the genotypes

selfed   selfed

selfed   selfed

selfed   selfed

selfed   selfed

selfed   selfed

selfed   selfed   selfed   selfed

Smaller beans within the group give rise to
offspring which are on average the same size
as those of larger ones. Selection does not act
within a single genotype.

Figure 1 Diagram summarizing Johannsen's experiment (based on Figure 11, Unit 19 of
S100 *Science: A Foundation Course*, The Open University Press).

At the beginning of the experiment the 25 smallest beans gave rise to progeny lines that
consistently produced beans with a lower average weight than the lines produced by the
heaviest beans. Within a line, however, the heaviest and the lightest beans gave rise to
progeny with essentially the same average weight.

QUESTION How does the answer to the previous question help to explain the results that Johannsen obtained with his inbred lines?

ANSWER Whenever selfing is practised, the chance of an individual being homozygous at a particular locus cannot be less than 0.5 and increases with each generation of selfing. After a few generations of selfing, therefore, individuals in a family are (a) likely to be of the same genotype and (b) will be homozygous at most loci. Consequently any phenotypic differences between individuals in the same inbred line must be the result of environmental effects, because the individuals do not differ genotypically at more than a few loci. Thus the differences between the weights of individual beans in Johannsen's inbred lines were the result of environmental effects and not of genetic differences.

9.1.2 Types of variation

Before going any further we must consider whether there are different types of variation. With the character of sex in mammals we can distinguish two forms— male and female. The determination of sex is complicated in detail, but in mammals it is based essentially on a chromosome switch mechanism, whereby an individual inheriting two X-chromosomes is female and an individual inheriting one X and one Y is male. Clearly there is a discontinuity between male and female. Intersexes do exist, but they are so rare that the general statement of discontinuity stands. When we study other characters like blood groups (Section 9.3) and haemoglobin, we find that there are also clear discontinuities between the different forms of the characters and that these differences are determined by *alleles or multiple alleles of one or a very small number of genes.*

QUESTION Can you recall another example of *discontinuous variation* which we showed in Unit 1 to have a genetic basis?

discontinuous variation

ANSWER Round as opposed to wrinkled seeds in *Pisum sativum.*

On the other hand, characters like seed weight in beans, plant height in tobacco, birth weight in man and butter fat content of milk from cows do not show discontinuities other than as a consequence of the units with which we weigh or measure these characters. In Unit 11 it is shown that the detailed study of *continuous variation* necessarily requires a quantitative approach using techniques of statistical analysis, but for the purposes of Units 9 and 10 it will be sufficient to say that continuous variation is determined partly by *alleles of genes at many loci, each having a small effect, and partly by effects of the environment.*

continuous variation

Other characters, though having a genetic basis that is essentially continuous, do not display a continuous range of phenotypic variation. For example, numerical, so-called *meristic characters* like petal number in celandines (*Ranunculus ficaria*) or bristle number in *Drosophila melanogaster* can only be expressed in incremental steps of whole petals or bristles. These characters show quantitative discontinuous variation, but because the underlying determination is continuous, the same statistical techniques are employed as in the analysis of variation that is more obviously continuous.

meristic character

9.1.3 Phenotypes and phenocopies

There is another difficulty that we must face. When a survey of the variation in a particular character has been made, there has always been a tendency to assume that the phenotype in one geographical location is determined by the same genetic system as in another location. Thus people of blood group A_1 in Japan are assumed to be blood group A_1 for exactly the same genetic reason as people in France are blood group A_1. If this assumption were not made, almost no research on human population genetics could be published because it would be almost impossible to obtain the necessary confirmatory evidence. This problem is elaborated in a different context in Unit 11.

As every schoolgirl knows these days, a woman who contracts German measles (rubella) during early pregnancy is more likely to produce a deaf child than a woman not suffering from the disease. Detailed studies have shown that the rubella virus is the agent that, among other effects, disrupts the development of the inner ear. Examination of pedigrees that include children suffering from rubella deafness indicates that the occurrence of deafness is sporadic and does not follow any pattern. This suggests that deafness is not heritable in these families. On the other hand there are pedigrees including deaf people which do show a clear pattern for the inheritance of the deafness.

QUESTION When we can identify a phenotype that is essentially environmentally determined and mimics a genetically determined form of that character, what do we call the non-heritable form?

ANSWER A phenocopy.

Rubella deafness is a phenocopy of genetically determined deafness. We must be aware of the possibility of phenocopying, because there could be horrifying confusion if we were studying the incidence of deafness in a population and assumed that deafness was always hereditary.

9.2 Variation and uniformity

9.2.1 Variation

At a very early stage in our lives, we learn to recognize people by integrating all their facial and other noticeable characteristics and subconsciously picking out the salient features that we use for recognition.

QUESTION Write down the names of ten relations, friends or work mates. *Without* checking, list the following characteristics against each name: eye (iris) colour; natural hair colour; whether the hair is naturally straight or curly.

ANSWER How well did you remember your friends? There is, of course, no right answer to the question. Its purpose was to show that people vary for all three of these characteristics, and incidentally that we often don't remember the information in a conscious way.

We could also have added the characteristics of sex, hair length and beard, but sex is an easy character to remember whereas hair length in both sexes and beards in men are matters of personal choice even though the potential is inborn.

Clearly some human beings have dark eyes and others have light eyes. We could extend the list of characters that show variation *ad nauseam*, but that would merely be collecting evidence of the same kind. A more interesting and penetrating question is to ask *why* do some people have dark eyes and others have light? There has in the past been a strong tendency to accept variation at face value and say that people have dark eyes because they have dark eyes. For various reasons that explanation has satisfied many people. But that answer is not very satisfactory when we find that different human populations have different frequencies of dark-eyed people. For example, the majority of Scandinavians have light eyes, the majority of Italians have dark. Why? Unfortunately the only evidence we have is circumstantial and relates to solar radiation, but there are many other examples of variation in man and other organisms that are more amenable to experimentation and analysis, and these examples nearly all have one feature in common. The individuals of one form are more likely to die before reproduction or have a lower fertility or fecundity than other individuals possessing a different form of the same character. Consequently the former make a smaller genetic contribution to the next generation.

In other words, there is *selection* against individuals of certain phenotypes and consequently against the genotypes determining those phenotypes. The evidence for these conclusions is discussed later in these Units and in Units 11–13. Some contrary evidence is also mentioned in later Units.

selection

399

9.2.2 Uniformity

Most of the characters we have discussed so far in these Units could be regarded as trivial. Indeed, earlier this century some people argued strongly that genetics dealt only with trivial variation and that the characters that were important were uniform and determined by cytoplasmic agents. It was argued that all mammals had nervous systems, circulatory systems and pentadactyl limbs based on the same plan and, although individual minor differences could be hereditary, the outline plan was not. On the contrary, we now consider that these systems are so fundamental to the normal development and functioning of the organism that with few exceptions any changes are lethal at a very early stage of development. There is, therefore, very strong selection for stability in development and it is not surprising that all members of a species, genus, family, etc. have many fundamental characters in common. We do not have the space to consider this problem further, but it is one you might think about as you reflect on what you have already studied and as you work through the rest of this Course.

9.3 Blood groups in humans. The ABO blood-group system

In 1900, while he was investigating the possibility of blood transfusion, Landsteiner discovered blood groups in humans. He spent the rest of his life working on blood groups and was the co-discoverer, in 1940, of the Rhesus blood-group system.

Subsequent work by many people in different parts of the world has shown that human populations are highly polymorphic for some 14 blood-group systems, of which the ABO and the Rhesus systems are clinically the most important. At an early stage it was shown that blood groups were inherited characteristics, although it was not until 1924 that Bernstein clarified the exact method of inheritance of the ABO blood-group system (see Table 1 for the alleles involved; you can deduce the dominance relationships from the genotypes). One important feature of blood groups is that environmental variation has no noticeable effect on the action of the genes involved, and for this reason they have been used in legal problems of identity and parentage (Unit 1).

During the First World War it was discovered that there were differences in the distribution of the ABO blood groups among American soldiers according to whether they were black or white, and subsequently blood groups have been used extensively in physical anthropology. The independent discovery of the Rhesus blood-group system by Levine and Stetson, and Landsteiner and Wiener, and its association with haemolytic disease of the newborn firmly established blood grouping as an important part of clinical and scientific medicine (Units 1 and 14).

What are blood groups? How are they detected? Analysis of whole blood reveals that in mammals there are two major components; one is cellular and the other fluid. The great majority of the cells are the red blood cells, which contain haemoglobin, and it is these that carry the blood-group specificity in the protein of their outer membranes. The red cells carry the *antigen* (S100, Unit 20) and it is this antigenic **antigen** protein that determines the blood group of a person. We shall ignore the white cells in the discussion that follows, but it is as well to be aware that there are also white-cell blood groups. The fluid component of the blood is *plasma* and when the clotting **plasma** agents, and some other substances, are removed it is then called *serum*. The serum **serum** contains the *antibodies* (S100, Unit 20); as far as the ABO and Rhesus blood-group **antibody** systems are concerned it is the red cells and the serum that we need to consider. A brief outline of blood grouping is presented in Race and Sanger's *Blood Groups in Man** and we cannot do better than quote the relevant section.

> The great bulk of the earlier blood group knowledge was gathered from the results of simple agglutination tests. A serum containing a known antibody is added to a saline **agglutination** suspension of red cells. If the cells carry the equivalent antigen they are agglutinated; if no agglutination occurs it is concluded that the cells lack the antigen. [See Fig. 2, facing p. 434.]

> The converse procedure is the identification of the type of antibody present in a serum by adding to the serum different samples of red cells containing various known combinations of antigens. By observing which mixtures show agglutination, the antibody or antibodies present in the serum can be determined by a process of elimination.

* From Race, R. R., and Sanger, R. (6th edn 1975) *Blood Groups in Man*, Blackwell.

A third situation, and much the most exciting, is that which arises when neither the antibody nor the antigen are known. The discovery of the blood group systems Lutheran, Kell, Lewis, Duffy, Kidd, etc., followed the finding of human sera which reacted with certain red cell samples irrespective of their content of known antigens. Other systems such as MN and P were discovered in a more deliberate way. Animals were injected with human red cells in the hope that some of the unknown blood group antigens on the red cell surface would stimulate the production of antibodies in the animal serum; and in the hope that when the immune serum came to be tested against different samples of human red cells only some samples would be found to contain the inciting antigen.

For the ABO blood-group system three antigens A_1, A_2, and B commonly occur and these are detected by three antibodies designated anti-A, anti-A_1 and anti-B, although you will also find the terminology α, α_1 and β antibodies used widely. Indeed the field of blood-grouping is rife with mixed and almost hopelessly confused nomenclature.

Although there are only three antibodies, they can be used to distinguish six different blood groups. Table 1 shows the reactions of the red-cell type to the antibodies and the frequencies with which the blood groups occur in southern England.

Table 1 The ABO blood groups: their genotypes, frequencies and reactions with specific antibodies

Antigen on red cells	Agglutination with antibodies			Frequencies* in southern England (in 3 459 persons)	Genotype
	anti-A	anti-A_1	anti-B		
A_1	√	√	—	34.81	$I^{A_1}I^{A_1}$, $I^{A_1}I^{A_2}$ or $I^{A_1}I^O$
A_2	√	—	—	9.89	$I^{A_2}I^{A_2}$ or $I^{A_2}I^O$
B	—	—	√	8.59	I^BI^B or I^BI^O
A_1B	√	√	√	2.62	$I^{A_1}I^B$
A_2B	√	—	√	0.64	$I^{A_2}I^B$
O	—	—	—	43.45	I^OI^O

* Data are from Ikin, E. W., Prior, A. M., Race, R. R., and Taylor, G. L. (1939). The distribution of the A_1A_2 BO blood groups in England. *Ann. Eugen.*, **9**, 409–11.

The situation is further complicated because the antibodies occur naturally in most humans (Table 2) and it is the presence of these antibodies which is particularly

Table 2 Naturally occurring antibodies in the serum of people of different ABO blood groups

Blood group (antigen on red cells)	Antibodies in serum
A_1	anti-B
A_2	anti-B, and anti-A_1 in 1–2% of A_2 people
B	anti-A
A_1B	none
A_2B	anti-A_1 in 26% of A_2B people
O	anti-A and anti-B

important as we shall see shortly. Figure 3* shows the distribution of the blood-group allele I^O in Europe. Clearly there is a *cline* (gradual change of frequency with respect to distance), with the frequency of I^O increasing from the south-east to the north-west. Why? We do not know why, but there have been numerous attempts to explain these distributions, some of which are considered below. **cline**

It was Ford who first pointed out that blood groups were examples of genetic polymorphism† and should, therefore, be subject to selection. Blood-group

* Figure 3 is from Mourant, A. E., Kopec, A. C., and Domaniewska-Sobczak, K. (2nd edn 1976) *The Distribution of the Human Blood Groups and Other Polymorphisms*, Oxford University Press.
† We have anticipated Ford's definition (p. 432) and the ways in which a genetic polymorphism can be maintained (p. 458). For the present take this sentence and the next on trust.

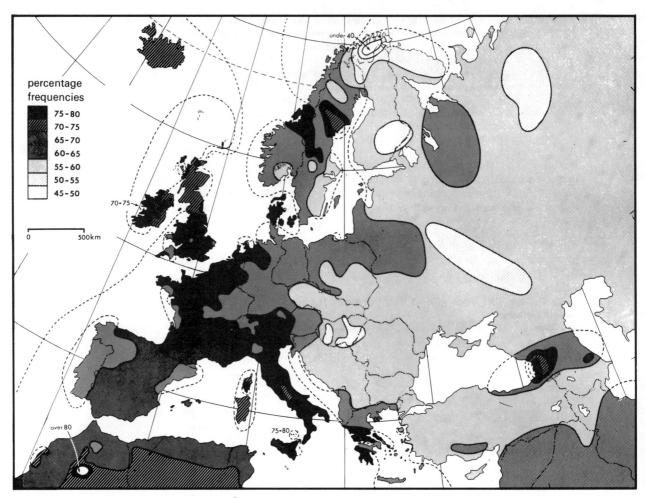

Figure 3 Distribution of the blood-group allele I^O in Europe.

frequencies appeared to remain almost constant from generation to generation and so the selective agents involved should be finely balanced. In the 1950s it was discovered that people of blood group O were 1.59 times as likely to suffer from duodenal ulcers as people of the other blood groups (see *STATS*, Section ST.4.7). Other work suggested that people of blood group A ($A_1 + A_2$) were 1.20 times as likely to suffer cancer of the stomach. The publication of these discoveries promoted a band-wagon effect with numerous laboratories investigating associations with infectious diseases on the one hand and with physiological disorders (cancer, ulceration, anaemias, diabetes, etc.) and congenital malformations (hydrocephalus, spina bifida, hare lip/cleft palate, etc.) on the other. There is little or no evidence for an association between the ABO blood groups and congenital disorders. It has been argued quite forcibly that many of the physiological conditions that we recognize today, occur at post-reproductive ages and may, therefore, have played only a minor role in the selective processes associated with the ABO blood-group system in the past. Some evidence from work by Jörgensen and Schwarz in West Germany is intriguing and is presented in Table 3. It is clear from these data that a healthy old age is more likely to be achieved if you are blood group O. Furthermore, people of blood group O are more likely to be physically active when they reach the age of 40. The significance of an excess of young men of blood group O volunteering for military service is difficult to interpret, but may be related to physical activity. One wonders whether there are more young men of blood group O playing professional football than we should expect!

Studies have been made on the fecundity of people of different blood groups and one large survey gave an excellent demonstration of where statistical techniques *could* lead the unwary astray. Several significant associations between blood group and fecundity were observed and the results looked as if they might help to explain the polymorphism. In the survey, the authors performed 397 χ^2 tests. Of these 23 (5.79 per cent) were significant at the 5 per cent level of probability and 9 (2.27 per cent) were significant at the 1 per cent level.

QUESTION Can one take the significant associations seriously?

ANSWER No. Marginally more tests were significant than would be expected if no true deviation from the null hypothesis were present. The observed significant χ^2s were attributable to chance alone. There is no real evidence, therefore, of differential fecundity. (See the treatment of type I and type II errors at the end of *STATS*, Section ST.7, p. 36.)

Table 3 Blood groups, health, and age[†]

Category of people	No. of Subjects	Controls	Ratio O:A	χ^2 for O:A (d.f. = 1)	Signif-icance
Healthy aged subjects (above 75) compared with German controls	521	81 985	1.597	23.219	* * *
Healthy aged subjects (above 75) compared with surgically treated aged patients	521	614	1.935	25.055	* * *
Athletes above 40 compared with German controls	340	81 985	1.34	9.937	* *
Volunteer soldiers compared with draftees	484	1 005	1.29	4.437	*

† Data are from Jörgensen, G., and Schwarz, G. (1968). Weitere Untersuchungen zur Frage der unterscheidlichen Selectionswertigkeit im ABO-Blutgruppensystem. *Hum. Genet.*, **5**, 254–60.

* $P < 0.05$; ** $P < 0.01$; *** $P < 0.001$.

The ratio O : A is obtained as follows. Suppose that the number of people in each of the four classes are as in the 2×2 table

	Blood group O	A
Subjects	f	g
Controls	h	l

then the ratio is calculated by $(f \times l)/(h \times g)$. The $\chi^2_{[1]}$ is a contingency chi square. (*STATS*, Section ST.4.7)

Up to this stage all the evidence we have been discussing appears to be negative. There is some strong positive evidence, but even this is still controversial. Vogel and his associates have suggested that the distributions of the various morphs of the ABO blood-group system, particularly in western Europe, reflect the distribution of two of the major epidemic diseases of the world: plague (*Pasteurella pestis*) and smallpox (variola). Perhaps their best evidence comes from a study of smallpox in rural Indian villages. These had no hospitals, or only very small ones, and no one had been vaccinated. They attempted complete ascertainment; that is, they tested everybody who had smallpox. The data for 1965 and 1966 are presented in Table 4. There is a significant excess of people of blood group A and AB who did not recover from the disease. Notice that Vogel used the sibs of affected individuals as his controls.

QUESTION Why, for this type of work, is it better to use sibs for controls rather than a random sample of the population from which the affected individuals come?

ANSWER There are basically two reasons.

(a) Apart from the genetic difference being studied the sibs of affected individuals are likely to be more similar genetically than other individuals. Consequently, this is the nearest we can get with humans to studying the effects of different alleles in the same genetic background.

(b) The environment of sibs is more likely to be more similar than that of non-sibs and, consequently, within each family the various individuals are more likely to have been exposed to the disease to similar extents. In this instance the data show that of the people who could have contracted the disease a marked excess of people of blood group A and AB did do so.

Table 4*   Differential effects of smallpox on sibs of varying ABO blood groups living in villages in India

1965	Blood group				
	A	B	O	AB	Total
Survived	43	30	22	2	97
Died	63	18	15	7	103
Controls (unaffected sibs of the above people)	38	66	84	12	200
1966					
Survived	42	36	4	19	101
Died	59	10	19	26	114
Controls (unaffected sibs of the above people)	19	101	76	11	207

* Data from Vogel, F., and Chakravartti, M. R. (1966). ABO blood groups and smallpox in a rural population of West Bengal and Bihar (India). *Hum. Genet.*, **3**, 166–80.

Comparisons A + AB : B + O

(a) for affected against
not affected sibs

$\chi^2_{[1]}(1965) = 42.25$ $P \ll 0.001$
$\chi^2_{[1]}(1966) = 121.57$ $P \ll 0.001$

(b) for those who survived against
those who died

$\chi^2_{[1]}(1965) = 8.65$ $0.01 > P > 0.001$
$\chi^2_{[1]}(1966) = 4.30$ $0.05 > P > 0.01$

Yates's correction for continuity has been used in these calculations (see *STATS*, Sections ST.4.2 and 4.7).

9.3.1 Summary of Section 9.3

Human populations are polymorphic for the various antigens of the ABO blood-group system. There is some evidence of differential survival and susceptibility to disease; but we have no adequate evidence to explain how the polymorphism is maintained and we cannot answer the question why there is a change in the frequency of the I^O allele from south-eastern to north-western Europe.

9.4   The quantitative approach to population genetics

9.4.1 Sickle-cell haemoglobin and allele frequencies

Having clarified the first steps in analysing variation in Sections 9.1 and 9.2, let us now consider an example in some detail so that we can pose in a precise way some questions that have been hinted at in previous Units. We shall return to sickle-cell anaemia again for our example, and as an *aide-mémoire* the points most relevant

Pedigree studies show that the HbA and HbS types of haemoglobin are determined by alleles Hb^A and Hb^S of the same gene. Heterozygotes Hb^AHb^S can be detected because their red cells become sickle shaped in the presence of ascorbic acid. (Ascorbic acid acts to lower the partial pressure of oxygen.) The red cells of Hb^SHb^S individuals assume the sickle shape naturally. If necessary, the haemoglobin of individuals whose red cells sickle could be studied by *electrophoresis** to check that the same form of the character is being detected in the different people involved. It is known that 80 per cent of the Hb^SHb^S homozygotes die before reproducing, whereas the heterozygotes have a 25 per cent better chance of

electrophoresis

* Electrophoresis is a technique for separating very small quantities of substances with similar structures. It involves the passing of an electric current through a gel or paper medium, along which the substances will migrate at speeds related to their electric charge, their size and shape. See p. 484, notes for Radio programme 9 and the demonstration in TV programme 12.

surviving attacks of the falciparum (sub-tertian) form of malaria than do $Hb^A Hb^A$ homozygotes. Clearly there is selection acting in favour of the $Hb^A Hb^S$ individuals in regions where this form of malaria is common.

> QUESTION Suppose you are studying a human population in Africa where the frequency of the Hb^S allele is 20 per cent, and malaria is suddenly eliminated as a major cause of death, which of the following would you expect to happen to the frequency of the Hb^S allele?
>
> (a) that it would remain roughly constant;
> (b) that it would decline at each generation at a predictable rate, until it is eliminated altogether;
> (c) that it would decline at each generation at a predictable rate to a level at which the rate of elimination equals the rate of production of new Hb^S mutants;
> (d) none of these?
>
> ANSWER With the inadequate information we have considered so far, we have no means, other than intuition, of concluding that the most reasonable answer to the question is (c).

Let us now see whether we can justify this conclusion. Clearly for this conclusion to be correct we must be able to estimate the 'predictable rate' of decline per generation. This implies that we *need to know the relationship between the frequency of alleles in one generation and the next.* We shall start with a basic situation and consider later (Section 9.4.4) the assumptions we have made. Suppose we have a large population of a diploid animal in which the frequencies of three possible genotypes at a hypothetical locus A are

$$A_1A_1 \quad A_1A_2 \quad A_2A_2 \quad \text{Total}$$
$$640 \quad\quad 320 \quad\quad 40 \quad\quad 1\,000$$

By dividing by the total (1 000) we can write the relative frequencies of the three genotypes as

$$0.64 \quad 0.32 \quad 0.04 \quad 1$$

By counting the alleles (two per individual) we can determine their relative frequency thus

$$A_1 = \frac{(2 \times 640) + 320}{2 \times 1\,000} = \frac{1\,600}{2\,000} = 0.80$$

$$A_2 = \frac{320 + (2 \times 40)}{2 \times 1\,000} = \frac{400}{2\,000} = 0.20$$

So, at the start the frequency of A_1 is 0.80 and the frequency of A_2 is 0.20. We now can determine what will happen when the population breeds. There are nine possible matings and these are shown in Table 5. Matings of the type $A_1A_1 \times A_1A_1$ will occur at a frequency $0.64 \times 0.64 = 0.409\,6$. (See *STATS*, Section ST.2.2 if you do not see why this is correct). All the progeny will be A_1A_1. Matings of the type $A_1A_2 \times A_1A_2$ will occur at a frequency $0.32 \times 0.32 = 0.102\,4$. Of the progeny $\frac{1}{4}$ will be A_1A_1, $\frac{1}{2}$ will be A_1A_2 and $\frac{1}{4}$ will be A_2A_2.

Table 5   The nine possible matings between individuals of three genotypes

♀	♂ Frequency	A_1A_1 0.64	A_1A_2 0.32	A_2A_2 0.04
A_1A_1	0.64	0.64×0.64	0.64×0.32	0.64×0.04
A_1A_2	0.32	0.32×0.64	0.32×0.32	0.32×0.04
A_2A_2	0.04	0.04×0.64	0.04×0.32	0.04×0.04

Thus by taking all possible combinations we can build Table 6, and we find that the frequency of the three genotypes is exactly the same among the progeny as it was among their parents. Similarly the allele frequencies have remained the same at 80 per cent A_1 and 20 per cent A_2. Thus it appears that these allele and genotype frequencies are at an *equilibrium*. We have shown numerically that the frequency of alleles in one generation is the same as in the previous generation.

equilibrium

Table 6 Mating and progeny frequencies

Parents ♀ ♂	Frequency of mating	Frequency of progeny		
		A_1A_1	A_1A_2	A_2A_2
$A_1A_1 \times A_1A_1$	0.409 6	0.409 6		
$A_1A_1 \times A_1A_2$	0.204 8	0.102 4	0.102 4	
$A_1A_1 \times A_2A_2$	0.025 6		0.025 6	
$A_1A_2 \times A_1A_1$	0.204 8	0.102 4	0.102 4	
$A_1A_2 \times A_1A_2$	0.102 4	0.025 6	0.051 2	0.025 6
$A_1A_2 \times A_2A_2$	0.012 8		0.006 4	0.006 4
$A_2A_2 \times A_1A_1$	0.025 6		0.025 6	
$A_2A_2 \times A_1A_2$	0.012 8		0.006 4	0.006 4
$A_2A_2 \times A_2A_2$	0.001 6			0.001 6
Totals	1.000 0	0.640 0	0.320 0	0.040 0

ITQ 1 Starting with the genotype frequencies

A_1A_1 A_1A_2 A_2A_2 Total
68%   24%   8%   100%

show that after one generation the genotype frequencies will become

64%   32%   4%

respectively.

We have now examined two situations. In the first one we found that the frequency of the alleles and the genotypes appeared to be at an equilibrium because they were the same in one generation as they were in the previous one. In the second case (ITQ 1) we started with different genotype frequencies, but with the same allele frequencies, and found that after one generation we attain the equilibrium frequencies for the genotypes.

Generation	Example 1			Example 2		
	A_1A_1	A_1A_2	A_2A_2	A_1A_1	A_1A_2	A_2A_2
0	0.64	0.32	0.04	0.68	0.24	0.08
1	0.64	0.32	0.04	0.64	0.32	0.04

9.4.2 The Hardy–Weinberg law

[A]*

For these examples we have used actual frequencies, yet what is of prime interest is whether this result can be generalized. We must start with the genotypes A_1A_1, A_1A_2 and A_2A_2 with respective frequencies P, $2Q$ and R, where $P + 2Q + R = 1$. (Experience shows that using $2Q$ rather than Q eases the algebra at later stages.)

Our table of genotype frequencies is

A_1A_1 A_1A_2 A_2A_2 Total
P $2Q$ R 1

For a generalized situation the frequency of A_1 must have any value between 0 and 1. Let us assume that the frequency of A_1 is p_0, where p_0 has a value between 0 and 1. The frequency of A_2 (which we shall call q_0) can only be $1 - p_0$: you will have to watch your ps and qs from now on! Remember that in general

$$p + q = 1 \qquad (1)$$

$p + q = 1$

QUESTION What are the values of p_0 and q_0 in terms of the genotype frequencies P, $2Q$ and R. (Use the counting method we used in Section 9.4.1. See also *STATS*, Section ST.12.3.)?

* See the Study guide for an explanation of this symbol.

ANSWER

$$p_0 = \frac{2P + 2Q}{2} = P + Q$$

$$q_0 = \frac{2Q + 2R}{2} = Q + R$$

Thus, $p_0 + q_0 = P + 2Q + R = 1$, so all is well.

You should now see why we chose to use $2Q$ rather than Q. If we had used Q, we would have halves in the equations, and so using $2Q$ avoids the introduction of fractions both here and in Table 7.

We can now construct the mating and progeny Table 7.

Table 7   Mating and progeny frequencies

Parents ♀ ♂	Frequency of mating	Frequency of progeny		
		A_1A_1	A_1A_2	A_2A_2
$A_1A_1 \times A_1A_1$	$P \times P$	P^2		
$A_1A_1 \times A_1A_2$	$P \times 2Q$	PQ	PQ	
$A_1A_1 \times A_2A_2$	$P \times R$		PR	
$A_1A_2 \times A_1A_1$	$2Q \times P$	PQ	PQ	
$A_1A_2 \times A_1A_2$	$2Q \times 2Q$	Q^2	$2Q^2$	Q^2
$A_1A_2 \times A_2A_2$	$2Q \times R$		QR	QR
$A_2A_2 \times A_1A_1$	$R \times P$		PR	
$A_2A_2 \times A_1A_2$	$R \times 2Q$		QR	QR
$A_2A_2 \times A_2A_2$	$R \times R$			R^2

Totals A_1A_1 $\quad P^2 + 2PQ + Q^2 = (P + Q)^2 = p_0^2$

$\quad\quad\quad A_1A_2$ $\quad 2PQ + 2Q^2 + 2PR + 2QR = 2(P + Q)(Q + R) = 2p_0 q_0$

$\quad\quad\quad A_2A_2$ $\quad Q^2 + 2QR + R^2 = (Q + R)^2 = q_0^2$

Thus we can express the genotype frequencies in one generation in terms of the allele frequencies in the previous generation

$$\begin{array}{cccc} A_1A_1 & A_1A_2 & A_2A_2 & \text{Total} \\ p_0^2 & 2p_0 q_0 & q_0^2 & 1 \end{array}$$

Note that $p^2 + 2pq + q^2 = (p + q)^2$. (See *STATS*, Section ST. 2.3). That is, the genotype frequencies correspond with the binomial expansion of the allele frequencies. Let us call the frequency of the A_1 allele in this progeny generation p_1. What is the value of p_1 in terms of p_0 ?

$$p_1 = \frac{2p_0^2 + 2p_0 q_0}{2} = p_0(p_0 + q_0) = p_0$$

Thus we have been able to show that the statement that the frequency of alleles remains constant from generation to generation is general for the type of situation we have been considering. This attainment of equilibrium is known as the *Hardy–Weinberg law* and is the jump-off point for all further discussion of population genetics.

Hardy–Weinberg law

In 1908 G. H. Hardy published a paper (*Science, N.Y.*, **28**, 49–50) called 'Mendelian proportions in a mixed population'. Quite independently and in the same year W. Weinberg's paper 'Über den Nachweis der Verebung beim Menschen' (*Jh. Ver. vaterl. Naturk. Württ.*, **64**, 368–82) was published. Both men have been acknowledged by the combination of their surnames to name the principle that they described. Hardy's paper is reproduced in full below, but we have changed his symbols to conform with the system used in these Units. We have highlighted some paragraphs by setting in bold type, partly because of the succinct way in which the important point was made, and partly because Hardy foresaw several implications of his model, which were clarified only 22 years later by Fisher and Haldane and Wright.

To the Editor of Science: I am reluctant to intrude in a discussion concerning matters of which I have no expert knowledge, and I should have expected the very simple point which I wish to make to have been familiar to biologists. However, some remarks of Mr Udny Yule, to which Mr R. C. Punnett has called my attention, suggest that it may still be worth making.

In the *Proceedings of the Royal Society of Medicine* (Vol. I., p. 165) Mr Yule is reported to have suggested, as a criticism of the Mendelian position, that if brachydactyly* is dominant 'in the course of time one would expect, in the absence of counteracting factors, to get three brachydactylous persons to one normal'.

It is not difficult to prove, however, that such an expectation would be quite groundless. Suppose that Aa is a pair of Mendelian characters, A being dominant, and that in any given generation the numbers of pure dominants (AA), heterozygotes (Aa), and pure recessives (aa) are as $P : 2Q : R$. Finally, suppose that the numbers are fairly large, so that the mating may be regarded as random, that the sexes are evenly distributed among the three varieties, and that all are equally fertile. A little mathematics of the multiplication-table type is enough to show that in the next generation the numbers will be as

$$(P + Q)^2 : 2(P + Q)(Q + R) : (Q + R)^2,$$

or as $P_1 : 2Q_1 : R_1$, say

The interesting question is—in what circumstances will this distribution be the same as that in the generation before? It is easy to see that the condition for this is $Q^2 = PR$. And since $Q_1^2 = P_1 R_1$, whatever the values of P, Q and R may be, the distribution will in any case continue unchanged after the second generation.

Suppose, to take a definite instance, that A is brachydactyly, and that we start from a population of pure brachydactylous and pure normal persons, say in the ratio of $1 : 10\,000$. Then $P = 1$, $Q = 0$, $R = 10,000$ and $P_1 = 1$, $Q_1 = 10,000$, $R_1 = 100,000,000$. If brachydactyly is dominant, the proportion of brachydactylous persons in the second generation is $20,001 : 100,020,001$, or practically $2 : 10,000$, twice that in the first generation; and this proportion will afterwards have no tendency whatever to increase. If, on the other hand, brachydactyly were recessive, the proportion in the second generation would be $1 : 100,020,001$, or practically $1 : 100,000,000$, and this proportion would afterwards have no tendency to decrease.

In a word, there is not the slightest foundation for the idea that a dominant character should show a tendency to spread over a whole population, or that a recessive should tend to die out.

I ought perhaps to add a few words on the effect of the small deviations from the theoretical proportions which will, of course, occur in every generation. Such a distribution as $P_1 : 2Q_1 : R_1$, which satisfies the condition $Q_1^2 = P_1 R_1$, we may call a *stable* distribution. In actual fact we shall obtain in the second generation not $P_1 : 2Q_1 : R_1$, but a slightly different distribution $P_1' : 2Q_1' : R_1'$, which is not 'stable'. This should, according to theory, give us in the third generation a 'stable' distribution $P_2 : 2Q_2 : R_2$, also differing slightly from $P_1 : 2Q_1 : R_1$; and so on. The sense in which the distribution $P_1 : 2Q_1 : R_1$ is 'stable' is this, that if we allow for the effect of casual deviations in any subsequent generation, we should, according to theory, obtain at the next generation a new 'stable' distribution differing but slightly from the original distribution.

I have, of course, considered only the very simplest hypotheses possible. Hypotheses other than that of purely random mating will give different results, and, of course, if, as appears to be the case sometimes, the character is not independent of that of sex, or has an influence on fertility, the whole question may be greatly complicated. But such complications seem to be irrelevant to the simple issue raised by Mr Yule's remarks.

P.S. I understand from Mr Punnett that he has submitted the substance of what I have said above to Mr Yule, and that the latter would accept it as a satisfactory answer to the difficulty that he raised. The 'stability' of the particular ratio $1 : 2 : 1$ is recognized by Professor Karl Pearson [*Phil. Trans. Roy. Soc.* (A), vol. 203, p. 60].

QUESTION Suppose we can distinguish the three phenotypes determined by the genotypes $B_1 B_1$, $B_1 B_2$ and $B_2 B_2$. How can we test whether the genotype frequencies in a sample of animals collected from a natural interbreeding population were in Hardy–Weinberg equilibrium. (*Hint*: a statistical test is needed—see *STATS*, Section ST.4.3.)?

ANSWER From the generalized Hardy–Weinberg law the expected frequencies of the three genotypes are p^2, $2pq$ and q^2, where p and q are the frequencies

* Brachydactyly means 'short fingers'.

of the B_1 and B_2 alleles, respectively. In the way we have done several times previously we can obtain the allele frequencies from the observed genotype frequencies. Substituting these values of p and q in the expression $p^2 : 2pq : q^2$ we obtain the expected genotype frequencies if the population is at equilibrium. We have to multiply to the total number of individuals in the sample to obtain the expected numbers of each genotype.

It then becomes a matter of calculating

$$\chi^2_{[d.f.]} = \sum_{i=1}^{n} \frac{(O_i - E_i)^2}{E_i}$$

ITQ 2 We have caught 500 butterflies from one interbreeding population and we find that 350 of them have spots on their hind wings and the rest have plain wings. By formal genetic analysis we find that 250 of the butterflies with spotted wings are heterozygous. Determine whether the genotype frequencies are in Hardy–Weinberg equilibrium.

When there is complete dominance, we cannot distinguish the heterozygote Aa from one of the homozygotes AA. In ITQ 2 the genotypes of these heterozygotes and homozygotes were distinguished by formal breeding and we could then count the alleles. Often, however, we cannot do this and so we have to resort to other methods. If we *assume* that the population is in Hardy–Weinberg equilibrium, we can then argue that the frequency of the distinguishable homozygote aa is the square of the frequency of the recessive allele. Thus we have

Genotype	AA	Aa	aa	Total
Frequency	p^2	$2pq$	q^2	1

The frequency of the a allele is

$$q = \sqrt{q^2} \qquad\qquad (2) \qquad\qquad q = \sqrt{q^2}$$

We can then estimate the frequencies of the AA and Aa genotypes.

QUESTION We have caught 400 individuals from another population of the butterflies mentioned in ITQ 2. Of these nine have plain wings. Assuming the population to be at equilibrium, estimate the frequency of the three genotypes in the sample?

ANSWER $q^2 = \frac{9}{400} = 0.022\,5$, and therefore

$$q = 0.15 \quad \text{and} \quad p = 1 - 0.15 = 0.85$$

The expected frequencies of the three genotypes are

	$SpSp$	$Spsp$	$spsp$	Total
Frequency	0.722 5	0.255 0	0.022 5	1
Expected numbers	289	102	9	400

Note Because we have assumed the Hardy–Weinberg law to hold, there is no point in testing whether it does hold!

Summary

1   Under conditions that will be elaborated in Section 9.4.4, the Hardy–Weinberg law shows that for one pair of alleles in a diploid organism the allele frequency will remain constant from generation to generation.

2   When the heterozygote between a pair of alleles is distinguishable from both homozygotes, the number of alleles of the two types can be calculated merely by counting the alleles.

3   On the other hand, when the heterozygote is indistinguishable from one of the homozygotes, the allele frequencies may be calculated by taking the square root of the frequency of the recessive homozygote.

4   Departures from the equilibrium genotype frequencies expected on the basis of the Hardy–Weinberg law can be detected by using the χ^2 statistic.

We could apply our basic model directly to the example of sickle-cell anaemia, but the arithmetic would be very tedious and experience shows that it is preferable to extend the model to allow for selection and so reduce the labour of calculation. Clearly the Hb^SHb^S individuals have a low survival rate; that is, they have low *fitness*. Fitness, in genetic terms, is related to the probability of leaving descendants many generations hence, and this obviously requires an individual to reach reproductive age and then reproduce. For the moment it matters little whether the selection occurs predominantly in adults, in juveniles or in reproductive capabilities, because the net effect will be the same. We shall discuss this further under the heading 'r and K selection' in Unit 13.

fitness

We are primarily interested in differential survival and in reproductive ability in a particular environment, and so we need the *relative fitness* of each genotype, compared with the others.

relative fitness

We take the fitness of the most efficient genotype as unity and compare the other genotypes with it. In what follows, we shall use $(1 - S)$ for fitness, where S is called the *selective coefficient*. Remembering from Section 9.4.1 that we are assuming that malaria has been eradicated and therefore that Hb^AHb^A and Hb^AHb^S individuals will be equally fit (that is, they have unit fitness), we now proceed to set up a modified Hardy–Weinberg model that allows for selection against Hb^SHb^S homozygotes.

selective coefficient

	Hb^AHb^A	Hb^AHb^S	Hb^SHb^S	Total
Frequency before selection	p^2	$2pq$	q^2	1
Numbers in a population of size N	Np^2	$2Npq$	Nq^2	N
Relative fitness	1	1	$1 - S$	
Numbers after selection	Np^2	$2Npq$	$Nq^2(1 - S)$	$Np^2 + 2Npq + Nq^2(1 - S)$

Let us assume that $S = 1$; that is, all Hb^SHb^S homozygotes die young, before reproducing. It follows that, after selection, the genotype frequencies will be

Hb^AHb^A	Hb^AHb^S	Hb^SHb^S	Total
Np^2	$2Npq$	0	$N(p^2 + 2pq)$

Now we are interested in the rate of change in the frequency of the Hb^S allele in successive generations. Before selection the frequency is q; let us call this q_0 so that we know the generation to which we are referring. After selection we obtain q_1 (subscript 1 because we are referring to generation 1). By counting alleles, q_1 in terms of q_0 is given by

$$q_1 = \frac{2Np_0q_0}{2N(p_0^2 + 2p_0q_0)}$$

We now proceed to simplify this expression. You should be able to follow the argument, but it is not necessary to remember the details.

By cancelling $2N$ we have

$$q_1 = \frac{p_0q_0}{p_0^2 + 2p_0q_0}$$

$$= \frac{p_0q_0}{p_0(p_0 + 2q_0)}$$

$$= \frac{q_0}{p_0 + 2q_0}$$

$$= \frac{q_0}{p_0 + q_0 + q_0}$$

$$q_1 = \frac{q_0}{1 + q_0} \tag{3}$$

$$q_1 = \frac{q_0}{1 + q_0}$$

Without working it out now, it can be shown that if we repeat this process for another generation, the relationship between the allele frequency in generation 2 (q_2) and that in generation 1 is

$$q_2 = \frac{q_1}{1 + q_1}$$

and generally it can be seen that

$$q_n = \frac{q_{n-1}}{1 + q_{n-1}}$$

What is of interest to us is the value of q_n in terms of q_0 because this will give us the over-all change in frequency over n generations. By substituting successive terms in the series we find, eventually, that

$$q_n = \frac{q_0}{1 + nq_0} \qquad (4)$$

$$q_n = \frac{q_0}{1 + nq_0}$$

This equation can be rearranged so that we can relate the number of generations to a desired change in allele frequency

$$1 + nq_0 = \frac{q_0}{q_n}$$

$$nq_0 = \frac{q_0 - q_n}{q_n}$$

$$n = \frac{q_0 - q_n}{q_0 q_n}$$

$$n = \frac{1}{q_n} - \frac{1}{q_0} \qquad (5)$$

$$n = \frac{1}{q_n} - \frac{1}{q_0}$$

Thus we can now use this equation to determine the number of generations (n) required to change the frequency of a recessive lethal allele by a specified amount. For Hb^S, we have $q_0 = 0.2$ (p. 405) and thus $1/q_0 = 5$.

The other term, $1/q_n$, is the vital one here because the smaller the value of q_n, the larger is the number of generations required to reach that low value. At the extreme when q_n is zero, n equals infinity and so the Hb^S allele can never be eliminated by selection alone! Thus, contrary to most preconceptions, selection alone is not capable of eliminating a recessive lethal mutation from a population. When the allele is rare it will necessarily occur only in heterozygotes and, being recessive, selection cannot influence the frequency of the allele.

But this is not the end of the story because we have still not obtained a means of calculating the change in frequency in successive generations, that is, the rate of change in the frequency of q.

QUESTION Starting with the equation $q_1 = q_0/(1 + q_0)$, show how we can calculate the change in frequency between successive generations.

ANSWER The change in the value of q from generation 0 to generation 1 can be written as $q_1 - q_0$. Using the equation $q_1 = q_0/(1 + q_0)$, we can subtract q_0 from both sides and obtain the change in the frequency of q (which is usually written as Δq) in terms of q_0. Thus

$$\Delta q = q_1 - q_0 = \frac{q_0}{1 + q_0} - q_0$$

Simplifying

$$\Delta q = \frac{q_0 - q_0(1 + q_0)}{1 + q_0}$$

$$\Delta q = -\frac{q_0^2}{1 + q_0}$$

What does this equation mean?

Remembering that q_0 can have any value between 0 and 1, we can determine Δq

411

for all permissible values of q. Leaving off the subscript we can write our equation as

$$\Delta q = - \frac{q^2}{1 + q} \qquad (6)$$

and construct a graph plotting Δq against q.

Inspection of this graph (Fig. 4) shows that the rate at which Δq changes depends on the value of q. In other words, there is an effect that depends on the frequency of the rare allele.

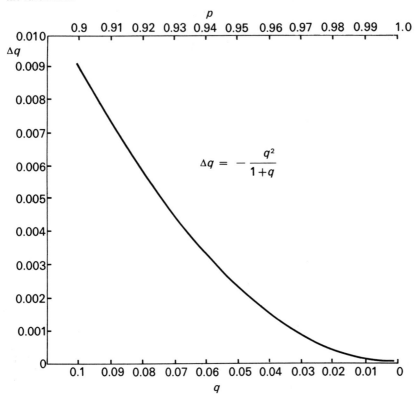

Figure 4 Graph showing the relationship between the change in allele frequency (Δq) and the allele frequency (q) for a recessive lethal allele. The upper scale shows the corresponding values of p. Remember that $\Delta q = -\Delta p$.

Unfortunately we have no means of studying directly the effect of the eradication of malaria on the frequency of the Hb^S allele, and so we cannot use this example to demonstrate that the theoretical model does have a real value in practice. This simple model of selection is not, however, just pie in the sky: an experiment with a lethal gene in a laboratory population of *D. melanogaster* reported by Wallace shows clearly a good correspondence between the results observed in an experiment and the results expected using equation 4, starting with $q_0 = 0.5$ (see Table 8 and Fig. 5).

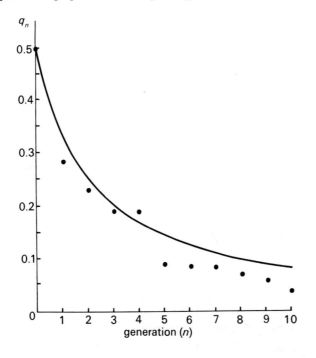

Figure 5 Graph showing the relationship between the allele frequency in a particular generation and the number of generations. The solid line represents the expected relationship assuming that q is the frequency of a recessive lethal allele which has an initial frequency of 0.5. The dots represent the actual frequencies that Wallace obtained.

Table 8   The values of the allele frequencies and number of generations used to construct the graph in Figure 5

Generation	Expected q_n	Observed q_n
0	0.500 0	0.500
1	0.333 3	0.284
2	0.250 0	0.232
3	0.200 0	0.189
4	0.166 6	0.188
5	0.142 9	0.090
6	0.125 0	0.085
7	0.111 1	0.082
8	0.100 0	0.065
9	0.090 9	0.054
10	0.083 3	0.041

Later (in Section 9.6.2) we reproduce part of the paper by Wallace, in which you will see that there is an even better interpretation of the observed frequencies, but we must now look again at the Hardy–Weinberg law.

Summary

1   A recessive lethal allele cannot be eliminated from a population by selection alone.

2   The rate at which a recessive lethal allele declines in frequency depends on its frequency at the time at which the rate is measured.

9.4.4   The assumptions behind the basic Hardy–Weinberg law

In Section 9.4.2 we developed the Hardy–Weinberg law from first principles, but there were several underlying assumptions that were not made explicit at the time. If any of these assumptions do not hold, then the genotype and/or allele frequencies are disturbed from their equilibrium values. We assumed, believe it or not, that:

1   there was random mating;
2   there was no differential selection;
3   there were no mutations;
4   there was no migration;
5   the population was large;
6   there were no effects of random processes;
7   it was possible to identify individuals;
8   generations were not overlapping;
9   the character we were studying was not sex linked.

These conditions are important and you should remember them.

We shall first expand on each point a little now, and then consider in depth, for the rest of these Units, the effect of some of these factors singly and together.

1   *Random mating* means that, allowing for the restraint of sex, an individual has the same chance of mating with any other individual in the breeding group. It is extremely unlikely that a population of plants or animals is strictly randomly mating, but for many purposes the effects of non-random mating can be ignored.

random mating

When mating between individuals of *different* phenotype with respect to a particular character predominates, it is called *disassortative* (or negatively *assortative*) *mating*. On the other hand, mating between individuals of like phenotype may occur preferentially, but we can often distinguish two situations: assortative mating and inbreeding. Mather, for example, has clarified the difference in the following way*:

assortative mating
dissassortative mating

* From Mather, K. (1973) *Genetical Structure of Populations*, Chapman & Hall, p. 31.

413

Assortative mating, which is the mating of phenotypically like individuals, is sometimes confused with inbreeding. Inbreeding, however, being the mating of relatives, always implies a particular relationship or set of relationships between mates in respect of all the genes that they carry and leads to homozygosis for all the genes in the genotype. Assortative mating does not do so. It will be effective only in relation to the genes affecting the character in respect of which the mating is assortative. Thus, for example, assortative mating in respect of leaf shape in plants might lead to homozygosis for genes affecting leaf-shape, but it would not do so for other genes affecting flower-colour or time of maturity or seed size, to mention but three other characters, except where the genes for these other characters were dragged along by linkage with the genes for the assortative character.

We shall not elaborate on the algebra of the situation, but merely point out that inbreeding, with all other things being equal, gives rise to changes in genotype frequency and not in allele frequency. It is also worth emphasizing that random mating is sometimes confused with the Hardy–Weinberg law itself; we hope that you will have a clearer understanding. (If you haven't, perhaps Radio programme 10 will help.)

2   Selection acts differentially on phenotypes and, therefore, changes both genotype frequencies and the frequency of alleles determining those genotype frequencies. (Section 9.6.1)

3   Mutation changes allele frequencies directly and clearly must also change genotype frequencies. (Section 9.5.3)

4   Migration is likely to change genotype frequencies, but the outcome depends on the genotypes and the numbers of the immigrant individuals. (Section 10.4.1)

5   The effect of population size would require too much explanation here, but it is considered later in the discussion of random processes. (Section 10.5)

6   *Genetic drift* is the name given to changes in allele frequency resulting from random processes. (Section 10.5.1)

genetic drift

7   With the majority of animals, individuals are clearly recognizable, they have a limited length of life and apart from certain exceptional circumstances they are all genetically different.

The models we have developed have all implicitly assumed that individuals can be distinguished and so they are applicable to most animal populations. Many plants, on the other hand, can maintain themselves for long periods without sexuality by means of what is loosely known as 'vegetative reproduction'. In consequence, dense clusters of plants of single genotypes (*clones*) may occur or there may be wide dispersal of a genotype in the form of a large number of isolated individuals (not considered further in this Course).

clone

8   The Hardy–Weinberg law assumes that mating occurs between individuals of the same generation. Clearly in humans, for example, several generations are alive and fertile at the same time and it is reasonable to think that this may have an effect on the frequency of alleles and genotypes in subsequent generations. If the population is at equilibrium, then *overlapping generations* are of no consequence. On the other hand a population not in equilibrium will reach equilibrium only when the last individual of the previous generation dies (not considered further in this Course).

overlapping generations

9   In the heterogametic sex (see Unit 2, Section 2.10), an individual can be neither heterozygous nor homozygous for alleles at a sex-linked locus. Consequently genotype frequencies must be different in the two sexes and this will affect the consequences of random mating (not considered further in this Course).

Although we have indicated that points 2 to 6 will be considered in more detail later in these Units, there are several other points that can be explained (but not proved) without the need to resort to algebra at this stage.

We have seen that a recessive lethal allele cannot be eliminated from a population as a result of selection alone. This allele could be lost by chance when it is rare, either because individuals possessing it failed to reproduce or zygotes possessing it failed to mature. And clearly, it is more likely that a rare allele will be lost by chance in a small population (less than 50 individuals) than in a large one. But new versions of this rare allele will be appearing in the population as a result of mutation, and random effects could *increase* the frequency of a rare allele rather than reduce its frequency. Thus, on the one hand, we have selection and chance effects reducing the frequency of an allele and, on the other, mutations and chance effects increasing its frequency. Is an equilibrium possible between the two processes? We cannot answer this question until we have considered mutation and selection in more detail. Mutation is the next topic.

Summary There are at least nine ways by which the Hardy–Weinberg equilibrium can be disturbed.

You should now attempt SAQ 1.

9.5 The dynamics of evolution: variation and mutation

To most people, evolution is a process of change and is, therefore, a dynamic process. In general this is true because, in the last analysis, evolution is recognized as having taken place when a change in allele frequency occurs. Sections 9.5, 9.6 and 9.7 of these Units are concerned with those forces that bring about changes in allele frequency and hence are concerned with the dynamics of evolution.

9.5.1 The types of discontinuous variation

In Section 9.1 we recognized two types of heritable variation. Characters showing continuous variation (for example, seed weight in beans) are determined partly by the cumulative effects of genes at many loci and partly by the influence of the environment during the course of development. Other characters show discontinuous variation (for example, HbA and HbS types of haemoglobin) and these are determined by the alleles of only one, or at the most a few, gene loci. We continue to concentrate on discontinuous characters, leaving a full discussion of the nature of continuous variation until Units 11 to 13.

We have already met throughout the earlier Units a great variety of characters showing discontinuous variation. To mention only a few examples: round versus wrinkled seeds in *P. sativum* and normal pigmentation versus albinism in humans Unit 1), adenine requirement versus adenine independence in *N. crassa*, white versus red eyes in *D. melanogaster* and resistance versus susceptibility of *E. coli* to lysis by T4 phage (Unit 2). Figure 6(i) shows some examples of mutation that have been found in wild populations of animals and plants. With the exception of albinism in humans, the examples discussed in previous Units were of experimental populations in which the particular phenotypes were maintained at high frequencies by appropriate breeding and selection techniques. In the present Units, however, we are interested in variation in natural populations, and a moment's reflection reveals that characters showing discontinuous variation fall into two broadly defined categories according to the frequencies with which they occur naturally.

Figure 6 faces pp. 418 and 419.

In the first category we have those characters where one form (the 'wild type') is shown by the majority of individuals and the variant occurs only very rarely. Albinism in humans and chlorophyll-deficient mutants (Fig. 6(ii)) in plants are examples of rare discontinuous variants. The second category contains those characters where two or more forms regularly occur together at moderately high frequencies within the same population as, for example, in the case of the ability, or not, of human individuals to taste phenylthiocarbamide (PTC). Such common variants are polymorphic (see footnote on p. 401). Remember that a particular character may only be considered to be 'rare' or 'polymorphic' in the context of a particular population at a given time. Thus the character HbA versus HbS haemoglobin is polymorphic among the indigenous populations of West Africa, but the *Hb*^{S} allele determining HbS haemoglobin is extremely rare among the Welsh. If malaria were completely eradicated from West Africa, we have predicted (Sections 9.4.1 and 9.4.3) that the status of the *Hb*^{S} allele would be reduced to that of a rare variant.

Having established this dichotomy of discontinuous variation we are naturally led to pose certain questions. For example, what is the origin of variation and how is it maintained? Why is some variation of the rare discontinuous type and other variation polymorphic? Can a rare variant become polymorphic, or vice versa, and if so how long does it take? It is to questions such as these that we shall address ourselves for the remainder of these Units. The answers, as we shall see, are not straightforward and some of them have not as yet been found.

9.5.2 Mutation as the origin of variation

If we regard evolution as a change in allele frequencies and, ultimately, as the loss of one form of a gene from a population and its replacement by the accumulation of another allele, we must pose questions regarding the origins of new forms of genes. We already know from Unit 2 that mutation is the primary source of new genetic variation, and in Section 2.11.1 mutants were classified as being advantageous or disadvantageous. Thus T4 phage resistance in *E. coli* may be classified as advantageous to the bacterium, but adenine requirement in *Neurospora*, achondroplasiac dwarfism and albinism in humans would be regarded as being disadvantageous. But such distinctions must be made in the context of the environment in which the mutant arises. Thus T4 phage resistance confers no advantage on *E. coli* in the absence of T4 phage (when it is probably even disadvantageous). On the other hand, the disadvantage associated with adenine requirement in *Neurospora* is greatly ameliorated when colonies are grown on medium supplemented with adenine.

The phenotypic effects of mutations to phage resistance or auxotrophy in *E. coli* or *Neurospora*, respectively, are immediately apparent because these organisms are haploids. The majority of higher organisms, however, are diploids and this introduces a further complication. We may safely regard both achondroplasia and albinism in humans as being disadvantageous, but remember that advantage, or otherwise, is a function of the phenotype of an individual, and not of its genotype. Thus the disadvantage associated with achondroplasia is manifested in an individual who carries just one copy of the mutant allele because achondroplasia is a dominant trait. On the other hand, because albinism is recessive, an individual carrying only one mutant gene suffers no disadvantage, and it is only in the recessive homozygote that the effect of the mutation becomes apparent. The distinction between dominant and recessive mutations is very important and we shall return to it later.

9.5.3 The effects of mutation on allele frequencies [C]

We have seen in Section 9.4 that, given certain assumptions, the Hardy–Weinberg law predicts that allele and genotype frequencies remain constant from generation to generation once an equilibrium has been established. These assumptions, therefore, define all the conditions necessary for no evolution to occur. By allowing these assumptions to be violated we can continue our enquiry into the evolutionary process.

The first assumption is that of random mating, but we have already stated, in Section 9.4.4, that non-random mating gives rise to changes in genotype frequencies but not in allele frequencies. In other words, the mating system does not itself affect the frequencies with which different alleles occur in different populations.

Another assumption is that of no mutation. We have already established mutation as the primary source of genetic variation, so it is logical to enquire to what extent the process of mutation can affect the subsequent distribution of variation. In a natural population mutant alleles arise in each generation, but the great majority of them are lost almost immediately. If a particular mutation of A_1 to A_2 has occurred just once, then, if we are considering a diploid organism, this new mutant allele must be present in a heterozygote, A_1A_2. All other individuals in the population will be homozygous A_1A_1 at that particular locus. If the mating system is such as to exclude selfing, one half of the progeny of the heterozygote will themselves be heterozygous carriers, and the other half will all be A_1A_1 homozygotes, which will not carry the new mutation. If the initial carrier has n progeny, then the probability that the mutation will not be passed on to the next generation is just $(\frac{1}{2})^n$. This means, given that n is not large, that a novel mutation is almost certain to be lost from a population within a few generations purely as a result of chance and, unless it enjoys an enormous selective advantage or happens to arise in a rapidly expanding population, it is unlikely to influence allele frequencies at that locus.

Mutations are not, however, one-off jobs, but recur regularly with a frequency that is predictable for a given locus. Consequently, in a large population there will always be a few representatives of a given mutant allele and not all of these will be lost by chance.

Given long enough we would expect mutant A_2 alleles to accumulate in the population and gradually replace A_1 alleles as a result of *mutation pressure*. But the situation is not as simple as this because A_2 alleles can also mutate back to A_1. Let us suppose

mutation pressure

that alleles A_1 and A_2 occur in the population with frequencies p_0 and q_0 in some initial generation. Let us further suppose that the mutation rate of A_1 to A_2, expressed as the fraction of A_1 alleles that mutate to A_2 between one generation and the next is μ and that the backward mutation rate of A_2 to A_1 is v. Since at the moment we are only concerned with the effect of *mutation* on allele frequencies, we shall assume that there is no advantage or disadvantage associated with the mutation, that is, that A_1 and A_2 are *selectively neutral* with respect to one another.

selectively neutral

So, in generation 0 in a population in which both A_1 and A_2 occur, A_1 has a frequency of p_0 and we expect a fraction μ of these A_1 alleles to mutate by the next generation. In generation 1, therefore, we expect μp_0 new A_2 alleles to have arisen by mutation from A_1. On the other hand we also expect $v q_0$ new A_1 alleles as a result of mutation from A_2. So on balance the new allele frequencies in generation 1 will be

$$p_1 = p_0 - \mu p_0 + v q_0$$

and

$$q_1 = q_0 + \mu p_0 - v q_0$$

and the change in allele frequencies between generations 0 and 1 will be

$$\Delta p = p_1 - p_0 = -\mu p_0 + v q_0$$

and

$$\Delta q = q_1 - q_0 = +\mu p_0 - v q_0$$

If in the initial generation the majority of alleles in the population are A_1 and only a few are A_2, then, provided the forward and backward mutation rates are of the same order of magnitude, there will be more A_1 alleles mutating to A_2 than vice versa. Thus, μp_0 will be greater than $v q_0$, and Δp will be negative whereas Δq will be positive. This means that the frequency of A_2 alleles will increase at the expense of A_1.

If initially, however, the majority of alleles are A_2, then the reverse argument holds, and the frequency of A_1 will increase and A_2 will decrease. It is clear that there must be an intermediate situation where the allele frequencies are such that the new alleles produced by mutation compensate for those lost and there will be no over-all change between generations. Such a situation implies that

$$p_n = p_{n-1} \quad \text{and} \quad q_n = q_{n-1}$$

or

$$\Delta p = \Delta q = 0$$

Then

$$\mu p = v q$$

Let the equilibrium values of p and q be p_e and q_e. Then

$$\mu p_e = v q_e$$

Adding $v p_e$ to each side

$$\mu p_e + v p_e = v q_e + v p_e$$
$$p_e(\mu + v) = v(p_e + q_e)$$
$$p_e(\mu + v) = v$$

Therefore

$$p_e = \frac{v}{\mu + v}$$

Similarly

$$q_e = \frac{\mu}{\mu + v}$$

Note that the mutation equilibrium values of p and q are quite independent of the initial allele frequencies and are determined solely by the relative magnitudes of the two opposed mutation rates. For example, if the mutation rates are equal, the equilibrium allele frequencies are 0.5. If the forward mutation rate is twice as large as the backward, then $\mu = 2v$ and $p_e = v/3v = \frac{1}{3}$ and $q_e = 2v/3v = \frac{2}{3}$.

But is an equilibrium ever achieved? We must first show that the allele frequencies always tend to move towards their equilibrium values. Now, as $p + q = 1$

$$\Delta q = \mu p - v q = \mu(1 - q) - v q$$
$$= -(\mu + v)q + \mu$$

But because $q_e = \mu/(\mu + v)$, we have

$$\Delta q = -(\mu + v)q + (\mu + v)q_e$$
$$\Delta q = -(\mu + v)(q - q_e)$$

QUESTION If $\mu = 3 \times 10^{-5}$ and $v = 1 \times 10^{-5}$ what is the equilibrium frequency, q_e, of the allele A_2?

ANSWER

$$q_e = \frac{\mu}{\mu + v} = \frac{3 \times 10^{-5}}{4 \times 10^{-5}} = 0.75$$

QUESTION Determine whether the frequency of the A_2 allele increases or decreases when its frequency in the preceding generation is (a) 0.70 and (b) 0.80.

ANSWER Inspection of the sign of Δq tells us whether the frequency increases (positive sign) or decreases (negative sign).

Now

$$\Delta q = -(\mu + v)(q - q_e)$$
$$= -(4 \times 10^{-5})(q - 0.75)$$

Therefore

(a) when $q = 0.70$

$$\Delta q = -(4 \times 10^{-5})(-0.05) = +2 \times 10^{-6}$$

and the frequency of A_2 increases;

(b) when $q = 0.80$

$$\Delta q = -(4 \times 10^{-5})(+0.05) = -2 \times 10^{-6}$$

and the frequency of A_2 decreases.

Because $(\mu + v)$ is always positive, Δq must be of opposite sign to $(q - q_e)$. If the frequency of A_2 is below the equilibrium value, that is if $q < q_e$, then $(q - q_e)$ is negative, and, therefore, Δq must be positive and the frequency of A_2 will increase. Conversely, if the frequency of A_2 is above the equilibrium value, that is if $q > q_e$, then $(q - q_e)$ is positive, and, therefore, Δq must be negative and the frequency of A_2 will fall. These arguments apply equally to the frequency of the A_1 allele. We see therefore that if the initial frequency is too large it will fall, and if it is too low it will rise. So when the allele frequency does not equal the equilibrium value it moves towards it, or when it is displaced away from the equilibrium frequency it tends to return. So, at least theoretically, our mutational equilibrium appears to be a perfectly respectable stable equilibrium. But how good is it in practice?

One of the necessary conditions for the Hardy–Weinberg law to hold was that the population size must be large. We have not, as yet, elaborated on the effects of population size, but we ask you to accept, for the moment, that the smaller the population the less likely it is that an equilibrium will be maintained by mutation alone. In fact, it can be shown that if both $4N\mu > 1$ and $4Nv > 1$ (where N is the population size), then the population is sufficiently large to allow an equilibrium to be maintained. This implies that if the mutation rates are of the order of 1×10^{-5} per generation the population size must be greater than 25 000 individuals.

It is also of interest to determine the length of time required for a given change in the frequency of an allele towards its theoretical equilibrium value. It can be shown that the number of generations required to change the allele frequency from p_0 (in some initial generation) to p_n (in generation n) is given by

$$n \approx \frac{\log_e(p_0 - p_e) - \log_e(p_n - p_e)}{\mu + v}$$

We have not derived this expression and we do not expect you to remember it, but the answers it gives are of interest. For example, if $\mu = 2 \times 10^{-5}$ and $v = 1 \times 10^{-5}$, then $p_e = 0.333$ and $q_e = 0.667$. Suppose we require the time to change from $p_0 = 0.9$ to $p_n = 0.5$. Substitution of these values in the above expression leads to the answer $n \approx 40\,800$ generations. Even on the evolutionary time-scale this is very long and it clearly demonstrates the weak nature of mutation pressure as an evolutionary process. Furthermore, we have been assuming that the two alleles, A_1

i(a)

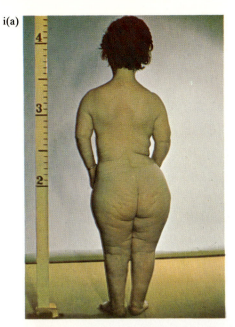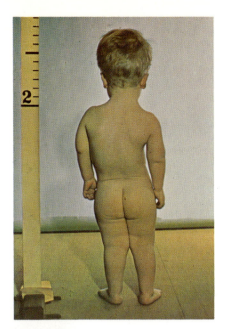

i(b)

i(c)

i(d)

Figure 6 (i)   Some rare mutant forms which have been found in natural populations of animals and plants and whose genetic basis is reasonably well understood.

(a)   Achondroplasia in humans. This is a congenital dominant condition which occurs at a frequency of, for example, approximately 1 in 14 000 births in Birmingham, England. The mutation rate in Denmark has been estimated as 4.3×10^{-5} births (see Unit 15, Section 15.4.2).

(b)   The *dohrnii* form of the currant moth *Abraxas grossulariata*. This character of reduced black markings on the wings is a sex-linked recessive and was one of the first to be studied (Doncaster and Raynor in 1906). The moths have been arranged in this photograph to show the pattern of inheritance.
One of the moths labelled as an F_2 male is, in fact, a female! This was merely an error in setting up the moths for the photograph.

(c)   White flowers in the bush vetch *Vicia sepium*. This recessive character is rare in the wild, but in one location in England it is much more common than we would expect. There is no explanation for this as yet.

(d)   A white flowered variant of the common primrose, *Primula vulgaris*. In 1973 two white flowered plants were found near Tamarton in Cornwall growing 1.2 km apart. In this region primroses are abundant. All 22 progeny produced by crossing these two plants had white flowers. Crossing one of the original white flowered plants with a normal yellow primrose gave 126 progeny all with yellow flowers.

ii

Figure 6 (ii) A recessive lethal mutation of a kind that may be observed not infrequently among germinating seeds of garden plants. This chlorophyll mutation has occurred in tobacco, *Nicotiana tabacum*. This species is self-fertile.

iii(a)

iii(b)

iii(c)

Figure 6 (iii) Variants that occur in wild populations at frequencies well above mutation rate (see Section 9.8.1).

(a) Dark and light keel petals in birdsfoot trefoil, *Lotus corniculatus*. The light form is recessive.

(b) A white flowered variant of the Spanish bluebell, *Endymion hispanicus*. The white form is recessive.

(c) Black spots, usually with yellow haloes, on the hind underwing of the meadow-brown butterfly, *Maniola jurtina*. This photograph is of the underside of a male with two spots per hind wing. Spot numbers can range from zero to five per wing. In England, nearly 70 per cent of the males have two spots. In Cornish females the zero-spot and two-spot forms are the most common (approx. 35–40 per cent each), whereas in the rest of England the zero-spot form is most common (approx. 60 per cent). (See E. B. Ford's book mentioned in the Bibliography for further details.)

iii(d)

(d) The radiate and non-radiate forms of the groundsel, *Senecio vulgaris*. It appears that a single allele pair T^r and T^n is involved, with heterozygotes having shorter ray florets than the $T^r T^r$ homozygote. The radiate form has been spreading in Britain during the past 100 years or so and currently (1976) it is the subject of study in several parts of the country. It has been suggested that the T^r allele has been introduced from the Oxford ragwort, *Senecio squalidus*, as a result of hybridization between native groundsel and *S. squalidus* that 'escaped' from the Oxford Botanic Garden and spread along the sides of railway tracks.

and A_2, are selectively neutral with respect to each other. It is extremely unlikely, however, that the environment will remain unchanged and that the state of neutrality will be maintained for such a long time. We may safely conclude, therefore, that mutation pressure cannot explain discontinuous polymorphic variation.

We usually take mutation rates to be of the order of 10^{-5} to 10^{-6} per generation, and somewhat less for lower organisms (for examples, see Unit 2). These are typical of the rates of mutation of wild-type (or normal) genes to their mutant alleles. But the back-mutation of mutant genes to their wild-type alleles is generally found to be much less frequent, or even nil, because those revertants to wild type which *are* observed are often mutations at other loci. Now we have shown that at equilibrium $\mu p_e = v q_e$, and, because μ is greater than v, this implies that p_e must be less than q_e. In other words the frequency of the wild-type gene must be less than its mutant allele! The consequence of this would be that if mutation were the only process in operation, populations would become full of strange and bizarre mutant forms. That this is not the case is patently obvious. A great variety of mutant genes do occur in populations, but they remain at a low frequency and account for the category of rare discontinuous variation. We clearly need to consider some process other than, or as well as, mutation. What could this other process be?

> QUESTION What other factors are likely to be of importance in changing allele frequencies.

> ANSWER Selection, migration, genetic drift (see Section 9.4.4 if you had difficulty in answering this question).

You should now attempt SAQ 2. (You will thereby make your own summary of Section 9.5.)

9.6 The dynamics of evolution: selection

9.6.1 A general model for selection   [B]

We must return to our original model of two alleles A_1 and A_2 alternative at locus A and consider the role of selection on the frequencies of the three possible genotypes A_1A_1, A_1A_2 and A_2A_2, and hence on the alleles A_1 and A_2. Selection can occur at four levels.

1 Individuals may die or otherwise be removed from a population before they have a chance to reproduce.

2 Individuals may reach maturity, but may not be able to reproduce; that is, they are genetically dead.

3 Individuals that are capable of reproduction may differ in their fecundity.

4 Selection may occur on the gametes themselves.

If we are interested in the fitness of a particular individual, the distinctions we have made above are important. When, however, we are concerned with the relation between the frequency of an allele in one generation and the next, the stage at which selection acts matters little for our present argument because the net effect will be the same. In other words, we are interested in the gametic contribution that all the individuals of a particular phenotype make to the next generation and not in the individual contributions.

Refer back to Section 9.4.3 and refresh your memory about the meaning of fitness. Let us suppose that the three genotypes A_1A_1, A_1A_2 and A_2A_2 have fitness $1 - S_1$, $1 - S_2$ and $1 - S_3$, respectively. We can construct the table:

	A_1A_1	A_1A_2	A_2A_2	Total
Genotype frequency	p^2	$2pq$	q^2	1
Fitness	$1 - S_1$	$1 - S_2$	$1 - S_3$	
Genotype frequency after selection	$p^2(1 - S_1)$	$2pq(1 - S_2)$	$q^2(1 - S_3)$	$\bar{w}$

where $\bar{w}$, termed the *mean fitness*, is given by $p^2(1 - S_1) + 2pq(1 - S_2) + q^2(1 - S_3)$. **mean fitness**
This can be simplified as follows. Expanding the factorized terms, we have

$$\bar{w} = p^2 - S_1 p^2 + 2pq - 2S_2 pq + q^2 - S_3 q^2$$

Since

$$p^2 + 2pq + q^2 = 1$$

$$\overline{w} = 1 - S_1 p^2 - 2S_2 pq - S_3 q^2 \tag{7}$$

The value of q before selection is given by

$$\frac{2pq + 2q^2}{2} = q(p + q) = q$$

using the fact that $p + q = 1$. The value of q after selection (q_1) is obtained from the genotype frequencies in the above table as

$$q_1 = \frac{2pq(1 - S_2) + 2q^2(1 - S_3)}{2\overline{w}}$$

$$= \frac{pq(1 - S_2) + q^2(1 - S_3)}{\overline{w}}$$

The change in q, Δq, is $q_1 - q$. Thus

$$\Delta q = \frac{pq(1 - S_2) + q^2(1 - S_3)}{\overline{w}} - q$$

The following procedure of rearrangement is tedious, but necessary, so that we can arrive at a more reasonable equation for Δq.

$$\Delta q = \frac{pq(1 - S_2) + q^2(1 - S_3) - \overline{w}q}{\overline{w}}$$

$$= \frac{pq(1 - S_2) + q^2(1 - S_3) - q(1 - S_1 p^2 - 2S_2 pq - S_3 q^2)}{\overline{w}}$$

$$= \frac{pq - S_2 pq + q^2 - S_3 q^2 - q + S_1 p^2 q + 2S_2 pq^2 + S_3 q^3}{\overline{w}}$$

$$= \frac{pq - S_2 pq(1 - 2q) + q^2 - S_3(q^2 - q^3) - q + S_1 p^2 q}{\overline{w}}$$

$$= \frac{pq - S_2 pq(p - q) + q(1 - p) - S_3[q^2 - q^2(1 - p)] - q + S_1 p^2 q}{\overline{w}}$$

$$= \frac{pq - S_2 p^2 q + S_2 pq^2 + q - pq - S_3 pq^2 - q + S_1 p^2 q}{\overline{w}}$$

$$= \frac{-S_2 p^2 q + S_2 pq^2 - S_3 pq^2 + S_1 p^2 q}{\overline{w}}$$

$$= \frac{pq[p(S_1 - S_2) - q(S_3 - S_2)]}{\overline{w}}$$

which we can also write as

$$\Delta q = \frac{pq}{\overline{w}}[p(S_1 - S_2) - q(S_3 - S_2)] \tag{8}$$

We have now obtained an expression for the change in allele frequency between generations, allowing for selection against the three genotypes and for different allele frequencies. Clearly this is a generalized model for selection.

ITQ 3 What does the Hardy–Weinberg law tell us about allele frequencies?

What is of particular interest now is whether the addition of selection into our discussion upsets the equilibrium and, if it does, whether there are any general principles that will help us to explain the role of selection in natural populations of animals and plants. We can have an equilibrium situation only when there is no change in allele frequency between successive generations, that is when $\Delta q = 0$.

So, under what conditions will Δq be zero? From equation 8 we can see that Δq will be zero when $pq = 0$, that is when $q = 1$ or 0, but these are only trivial equilibria because they imply that one or other allele is fixed in the population. In other words in the absence of mutation or migration there can be *no* further change. On the other hand, Δq will also be zero when $p(S_1 - S_2) = q(S_3 - S_2)$.

Rearranging this equation with respect to q we have

$$(1 - q)(S_1 - S_2) = q(S_3 - S_2)$$
$$S_1 - S_1q - S_2 + S_2q = S_3q - S_2q$$
$$2S_2q - S_1q - S_3q = S_2 - S_1$$

and thus at equilibrium we can write

$$q_e = \frac{S_2 - S_1}{2S_2 - S_1 - S_3} \qquad (9)$$

$$q_e = \frac{S_2 - S_1}{2S_2 - S_1 - S_3}$$

This means that the allele frequency at equilibrium is a function of the selective coefficients and *nothing else*.

Is this equilibrium stable? It requires only a very small change in the relative values of S_2 and S_1 for there to be an immediate effect on the frequency of the A_2 allele. If q is neither 0 nor 1, it can only have a value between 0 and 1 and clearly will be positive.

A minor adjustment to the denominator of the equilibrium equation will help clarify the following arguments:

$$q_e = \frac{S_2 - S_1}{(S_2 - S_1) + (S_2 - S_3)}$$

Firstly, as q is always less than one, the numerator must always be smaller than the denominator, and consequently $S_2 - S_1$ and $S_2 - S_3$ must both be positive or both be negative. When S_2 is larger than S_1, S_2 must also be necessarily larger than S_3. Similarly, when S_2 is smaller than S_1, S_2 must also be smaller than S_3.

We can have, therefore, two situations:

(a) heterozygous advantage, where $S_1 > S_2 < S_3$, which, as we shall see shortly, is a *stable* equilibrium, or

(b) heterozygous disadvantage, where $S_1 < S_2 > S_3$, which is unstable and leads to fixation of one or the other allele.

9.6.2 Special cases [B]

We shall return to equilibrium situations shortly, but this is the place to consider four special cases.

For three of these we take A_1 as an advantageous allele (that is, $S_1 = 0$) whereas for the fourth the heterozygote is at an advantage (that is, $S_2 = 0$). In the table below, whenever the fitness value is unity, we have taken the corresponding value of S as zero. The letter h allows for variation in the dominance relationship between A_1 and A_2. When $h = 0$, A_1 is dominant and when $h = 1$, A_2 is dominant as far as fitness is concerned. The reason why the selective coefficients do not carry subscripts in the first three of these special cases should become apparent in the subsequent discussion.

Special conditions	A_1A_1	A_1A_2	A_2A_2
(a) A_1 is dominant for fitness	1	1	$1 - S$
(b) A_2 is dominant for fitness	1	$1 - S$	$1 - S$
(c) The heterozygote has intermediate fitness	1	$1 - Sh$	$1 - S$
(d) The heterozygote has an advantage	$1 - S_1$	1	$1 - S_3$

We can now substitute for S_1, S_2 and S_3 in equations 7 and 8

$$\Delta q = \frac{pq}{\bar{w}} [p(S_1 - S_2) - q(S_3 - S_2)] \qquad (8)$$

$$\bar{w} = 1 - S_1p^2 - 2S_2pq - S_3q^2 \qquad (7)$$

421

Special case (a)

When $S_1 = S_2 = 0$ and $S_3 = S$

$$\Delta q = \frac{pq(-Sq)}{1 - Sq^2}$$

$$= -\frac{Spq^2}{1 - Sq^2} \tag{10}$$

$$\Delta q = -\frac{Spq^2}{1 - Sq^2}$$

Clearly there can only be an equilibrium when $q = 0$ or 1 ($p = 0$), because for all other possible values of q, Δq is negative and not zero. Furthermore, we can use this expression to gain some idea of the rate of decline in the frequency of the A_2 allele. When q is small, q^2 is very small and p is approximately equal to one. Therefore, the rarer the A_2 allele, the slower is the decrease in its frequency.

Special case (b)

When $S_1 = 0$ and $S_2 = S_3 = S$

$$\Delta q = \frac{pq[-Sp]}{1 - 2Spq - Sq^2}$$

$$= -\frac{Sp^2q}{1 - S(2pq + q^2)} \tag{11}$$

$$\Delta q = -\frac{Sp^2q}{1 - S(2pq + q^2)}$$

Again equilibrium is reached only when $q = 0$ or 1.

Special case (c)

When $S_1 = 0$, $S_2 = Sh$ and $S_3 = S$, the heterozygote is of intermediate fitness. The value of h can vary between 1 and 0 and therefore it allows for varying degrees of dominance for fitness.

$$\Delta q = \frac{pq(-Shp - Sq + Shq)}{1 - 2Shpq - Sq^2}$$

$$= -\frac{Spq[h(p - q) + q]}{1 - 2Shpq - Sq^2} \tag{12}$$

$$\Delta q = -\frac{Spq[h(p - q) + q]}{1 - 2Shpq - Sq^2}$$

At equilibrium we can have $q = 0$ or 1, or alternatively

$$h(p - q) + q = 0$$
$$hp - hq + q = 0$$

Therefore

$$q = -\frac{hp}{1 - h}$$

Clearly, whatever the values of h and p between their limits of 0 and 1, q will be negative! But we know that q must also have a value between 0 and 1, and therefore an equilibrium cannot exist unless $q = 0$ or 1.

QUESTION So far there have been no examples in this Section. Can you suggest why?

ANSWER The generalized model did not relate to any specific situation, but was the basis for deriving special cases. Not one of the three special cases achieves a stable equilibrium and so any examples are likely to be very transitory. We have already considered the recessive lethal situation in Section 9.4.3, but we can now consider Wallace's paper*.

In his *Introduction to Quantitative Genetics* Falconer (1960, p. 34) illustrates the elimination of a deleterious gene from a population by the use of a sex-linked mutant. He explains his choice by saying that there appears to be no well-documented example for an autosomal gene.

While re-examining notes on experimental populations of *Drosophila melanogaster* studied at Cold Spring Harbor, I found unpublished data that may have some use as a classroom example. They concern the elimination, from a population, of a lethal [*lt*-lethal] for which the parental flies were all heterozygous... Generations in this study were discreet, so uncertainties regarding generation time in a continuously breeding population are not involved in the interpretation of the data.

* Extracts from Wallace, B. (1963). The elimination of an autosomal lethal from an experimental population of *D. melanogaster. American Naturalist*, **97**, 65–6.

Table I The elimination of *lt*-lethal from a population. 'Obs' equals observed frequency, '*n*' equals the number of chromosomes tested (equals twice the number of flies tested), 'Exp. 1' equals frequencies expected for a recessive lethal, and 'Exp. 2' equals frequencies expected for a lethal that lowers the fitness of heterozygotes by 10 per cent.

Generation	n	Obs.	Exp. 1	Exp. 2
0	—	0.500	—	—
1	454	0.284	0.333	0.32
2	194	0.232	0.250	0.23
3	212	0.189	0.200	0.17
4	260	0.188	0.167	0.13
5	290	0.090	0.143	0.11
6	398	0.085	0.125	0.09
7	366	0.082	0.111	0.08
8	382	0.065	0.100	0.07
9	388	0.054	0.091	0.06
10	394	0.041	0.083	0.05

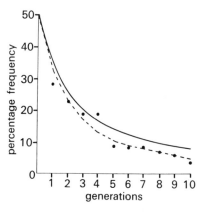

Figure 1 The observed decrease in the frequency of *lt*-lethal in successive generations. The solid line represents the decrease expected for a completely recessive lethal; the broken line represents the decrease in frequency for a lethal that lowers the fitness of heterozygotes by 10 per cent.

In the table [Table I] we have listed the observed frequencies of the lethal in successive generations together with frequencies expected in the case of a complete recessive and of a lethal that lowers the fitness of its heterozygous carriers by 10 per cent. The observed frequencies follow fairly well those based on the lower fitness of lethal heterozygotes (figure I). There is a suggestion that the selection pattern changed between the fourth and fifth generations but certainly this is no more than a suggestion.

Starting with $q = 0.5$ the Exp. 1 frequency is obtained by substituting successive values of q in the equation $q_1 = q_0/(1 + q_0)$, which we described in Section 9.4.3. The Exp. 2 frequency is calculated from the equation

$$q_1 = \frac{2pq(1 - S_2) + 2q^2(1 - S_3)}{2\bar{w}}$$

which when we have substituted for $\bar{w}$, becomes

$$q_1 = \frac{pq(1 - S_2) + q^2(1 - S_3)}{1 - S_1p^2 - 2S_2pq - S_3q^2}$$

But this is an unwieldy equation to use. Let us see whether we can make life easier for ourselves and so reduce the likelihood of error. Check that when $S_1 = 0, S_3 = 1$ and $S_2 = S_3h = h$, the equation for q_1 above becomes

$$q_1 = \frac{pq(1 - h)}{1 - 2pqh - q^2}$$

Substituting $h = 0.1$ and $1 - h = 0.9$, the equation becomes

$$q_1 = \frac{(1 - q)q \times 0.9}{1 - 2(1 - q)q \times 0.1 - q^2}$$

$$= \frac{0.9(q - q^2)}{1 - 0.2q - 0.8q^2}$$

And this manipulation makes the subsequent arithmetic for calculating the Exp. 2 value much easier.

In Figure 5 we found a good correspondence between the results observed in the experiment and the results expected with a recessive lethal allele. We have now seen that an even better relationship exists between the observed and expected frequencies when, in addition, we assume that the heterozygote has a 10 per cent lowered fitness.

Special case (d)

When $S_2 = 0, 0 < S_1 \leqslant 1$ and $0 < S_3 \leqslant 1$

$$\Delta q = \frac{pq(S_1p - S_3q)}{1 - S_1p^2 - S_3q^2} \qquad (13)$$

$$\Delta q = \frac{pq(S_1p - S_3q)}{1 - S_1p^2 - S_3q^2}$$

Again an equilibrium is possible when $q = 0$ or 1, but there is also an equilibrium when $S_1p - S_3q = 0$, that is, when

$$S_1p = S_3q$$

QUESTION Is this a stable equilibrium?

ANSWER Suppose the A_2 allele increases in frequency. Then q increases and $S_1p - S_3q$ has a negative value. Consequently Δq is negative and the frequency of the A_2 allele declines. On the other hand, when the A_2 allele decreases, $S_1p - S_3q$ has a positive value. Δq will be positive and the frequency of the A_2 allele will increase.

This equilibrium is maintained by the balance between the selective forces S_1 and S_3 acting on the homozygotes. Our original premise in this special case was that the heterozygote was at a selective advantage and clearly it is possible by these means to maintain both alleles in a population at a frequency above mutation rate. This situation has been described as genetic polymorphism.

The only example of a genetic polymorphism that is *known* to be maintained by a balance of selective forces is our perennial example of sickle-cell anaemia*. On the other hand, many studies have been made of genetic polymorphisms and some of the selective agents involved have been identified. Unfortunately it has not been possible to estimate the precise effect of any of these agents in natural populations although it has been possible to do so in the unnatural environment of the laboratory. We have similar problems with examples of selection in action. It is easy to show the effects of selection in the laboratory, but it is very difficult to *prove* selection in the wild even though selection is a likely interpretation of what has been found.

Summary

This Section can be summed up by equations 7–13.

9.6.3 The conflict between ecological and population genetics

Various people during the past 20 years or so have made rather depressing statements about the utility of the mathematical theories of evolution based on Mendelian and post-Mendelian genetics, essentially because the various parameters involved concern selective advantage, mutation rates and population size, most of which are still too inaccurately known to enable quantitative predictions to be made or verified. We have very little information on precisely why certain characters are selected under a given set of environmental conditions. Why are we so woefully ignorant?

Let us explore the reasons why it has proved difficult to match up these models with the real world of natural populations of animals and plants.

> ITQ 4 Following on from ITQ 2, suppose we catch 500 butterflies from another population and find the following distribution of genotypes
>
> | SpSp | Spsp | spsp | Total |
> |------|------|------|-------|
> | 80 | 270 | 150 | 500 |
>
> Are these genotype frequencies in Hardy–Weinberg equilibrium?

After working through the χ^2 test, we come to the conclusion that it is very unlikely that these genotype frequencies are in Hardy–Weinberg equilibrium. The data themselves cannot tell us why, but we are justified in asking the question. Clearly we must conclude that one of the conditions we outlined in Section 9.4.4 does not hold. We can recognize individuals, overlapping generations are unlikely, and, since we have a sample of 500 individuals, the population can be regarded as 'large'. We know nothing about random mating, migration, mutation or selection in the population, but we can argue about them. Both the migration and mutation rates would have to be unusually large if either is to account alone for the disturbance from Hardy–Weinberg equilibrium. Note that the heterozygote is the commonest genotype: do we have an example of heterozygous advantage? It is clear that both the *Sp* and *sp* alleles occur in the population at well above mutation rate, and so we are justified in concluding, initially, that we are studying a genetic polymorphism. Certainly this is a good jump-off point for further investigation, but we need to know the way selection is acting. Up to this stage we have relied almost entirely

* Recent studies on warfarin resistance in rats also strongly suggest a situation of heterozygous advantage.

on the model; indeed, without the model we would have no signposts for our investigation. But the model cannot tell us what to do next! This can only come from going back to the original population. The habitat of the butterfly needs to be examined, so does its life history, its food plants, its parasites, its fecundity, let alone its behaviour and distribution. And, of course, none of this information can be collected overnight.

Before we place the model on the shelf—to await verification or otherwise—there is another way it can help in our investigation. Suppose we visit the original population again and find that the frequency of the three genotypes remains approximately stable for several years. Allowing that we have also confirmed that the genetic system involved depends on a pair of alleles—that is, that we are now justified in concluding that we *are* studying a genetic polymorphism—we can estimate the selection acting on the polymorphism. We must assume random mating and that progeny are initially produced in binomial proportions, and then estimate what selection would be involved in changing the relative frequencies to those observed in our sample.

Taking the numbers from ITQ 4 we argue

	SpSp	Spsp	spsp	Total
Before selection	92.45	245.10	162.45	500
After selection	80	270	150	500

Taking the fitness of the *Spsp* genotype as unity we calculate the relative fitness of *SpSp* as the ratio of the fitness of the *SpSp* homozygote to the *Spsp* heterozygote, thus

$$\frac{80/92.45}{270/245.10}$$

This can be rewritten as

$$\frac{80}{92.45} \times \frac{245.10}{270} = 0.785\ 5$$

Similarly the relative fitness of *spsp* is

$$\frac{150}{162.45} \times \frac{245.10}{270} = 0.838\ 2$$

Thus in our model

	SpSp	Spsp	spsp
	$1 - S_1$	1	$1 - S_3$
We write	$1 - 0.214\ 5$	1	$1 - 0.161\ 8$
or	0.785 5	1	0.838 2

To check that we have done our sums right we can start with

	SpSp	Spsp	spsp	Total
Before selection	92.45	245.10	162.45	500
Fitness	$1 - 0.214\ 5$	1	$1 - 0.161\ 8$	
After selection	72.619 5	245.10	136.165 6	453.885 1

We now need to convert the numbers proportionately so that they sum to 500. Therefore we multiply by 500/453.885 1:

80	270	150	500

These numbers are the same as those observed.

All this is splendid until you realize that the model, and the estimates of selection obtained, tells us only what is going on. It does not, and indeed cannot, tell us why. What are the agents of selection, how do they work, are they evenly distributed over the population, are they constant or are they rhythmical—that is, diurnal or seasonal—do they interact? It is little wonder that the ecological geneticist argues that the models are not of great help. On the other hand, the models can often suggest what we need to study and certainly delineate the terms of reference within which the answers must be sought.

Finally notice that a genetic polymorphism of the type we explored just now with the butterflies is an equilibrium situation; that is, there is a tendency to maintain a *status quo*. An evolutionary process is involved, but the effect is static! Before dealing with the statics of evolution any further we must examine some examples of evolution in action.

You should now attempt SAQ 3.

9.7 The dynamics of evolution: evolution in action

9.7.1 Industrial melanism in the peppered moth

Geneticists have now studied numerous examples of evolution in action and, of these, antibiotic resistance in bacteria, DDT resistance in insects, warfarin resistance in rats and myxomatosis resistance in rabbits were outlined in S100, Section 19.6.1. The example that created the most impact at the time, and is still the most spectacular, was Kettlewell's study of the melanic (dark) form of *Biston betularia* (the peppered moth). Because the melanic form is commoner in industrial areas than elsewhere, the phenomenon has been named *industrial melanism*. It is only since 1955 that the role of selection in the evolution and spread of melanism has been understood. The two stumbling blocks in earlier studies were arguments over the heritable nature of melanism and whether birds ate butterflies and moths. Some rather sloppy work in the 1920s suggested that melanism in several species could be induced by feeding the larvae on polluted food. Subsequently these Buffonian*-type explanations were refuted, and Ford suggested that in industrial areas the melanic forms were at an advantage because they were well camouflaged (that is they were *cryptic*) on dark (sooty) tree trunks (Fig. 7), whereas the typical forms were obvious to potential predators. By experimentation, using families in which black and pale forms segregated, Ford also demonstrated that under conditions of semi-starvation the larvae of the melanic form of the moth *Cleora repandata* survived better than the typical form. In addition to crypsis, therefore, the melanic form would be at an advantage in areas where food was limiting.

industrial melanism

cryptic

Figure 7 Industrial melanism in the peppered moth *Biston betularia*. The typical and melanic forms on (a) lichen-covered bark, and (b) lichen-free bark coated with soot.

(a)

(b)

* It was Buffon, and not Lamarck, who first suggested the inheritance of acquired characters. Lamarck's great contribution is that he championed the theory of evolution *with reasoned argument* invoking 'use and disuse', a reasonable conclusion to draw from the facts available to him.

This work presupposed predation, indeed selective predation, but at that time (up to about 1955), apart from Ford and Hale-Carpenter, very few people believed that birds ate butterflies and moths. This attitude was eminently reasonable because a bird would have to suffer the sharp and unpalatable wings in return for the small, yet tasty body of the insect. What Ford and Hale-Carpenter interpreted as beak marks on the wings (Fig. 8), was attributed by other people to accidental damage caused when the insect was blown about in the wind.

Figure 8 Damage to the wings of butterflies. Clear beak marks and major wing damage can be seen.

Apart from his breeding work with *B. betularia*, which confirmed monohybrid inheritance and that the melanic form was completely dominant to the peppered form, Kettlewell made two vitally important discoveries at an early stage of his study. Firstly he was able to show that when he released marked moths in Dorset and in Birmingham he recovered a greater proportion of the light form than of the dark form in Dorset and a greater proportion of the dark than the light form in Birmingham. Secondly he was able to demonstrate, with the help of Tinbergen and his ciné camera, that several species of birds were avid eaters of the moths. The wings of the moths were no deterrent to the birds after all.

Because this work made fundamental contributions to the study of selection in action, and subsequent studies have also clarified the bases of some arguments about the phenomenon of dominance, it is easy to justify why we should discuss the experiments of Kettlewell and his colleagues in a little more detail.

The peppered moth can be trapped using mercury vapour light traps or by 'assembling' to virgin females. Experience showed that females were rarely caught using light traps and clearly only males will be attracted to virgin females, and so the experiments of marking, releasing and recapturing moths are only relevant to males.

Kettlewell bred a stock of several thousand pupae of *B. betularia*. The emerging male adults were marked with quick-drying cellulose paint and released in the wild. He chose two locations for the work, Deanend Wood in Dorset and the Christopher Cadbury Reserve, Rubery, near Birmingham. In Birmingham, where the tree-trunks were blackened by soot, lichens were totally absent and leaf washings in late summer revealed large amounts of pollution fall-out, the melanic form was common. In the rural setting of Dorset the tree-trunks were well covered with lichen and the typical form of the moth was well camouflaged (see Fig. 7).

Known numbers of marked individuals of each form were released and the traps were set up the following night. Table 9 shows the numbers of males released and recaptured in three experiments carried out in 1953 and 1955.

If we take just the Dorset experiment, it is clear that a greater proportion of the typical form (*typica*) is recaptured than of the melanic form (*carbonaria*), but there are several possible explanations for this. The life-span of the two forms may be different, melanics may be attracted to light or to virgin females to a different extent from the typicals, the two forms may wander or migrate to different extents, or they may be subjected to differential predation. Taking the Dorset experiment alone

Table 9*   **Recovery of the typical and melanic forms of *Biston betularia* released into an unpolluted area in Dorset and one polluted by smoke near Birmingham**

		Typical	Melanic	Total
Deanend Wood	Frequency in the natural population	297	0	297
Dorset	Marked released	496	473	969
1955	Marked recaptured	62	30	92
	% of releases recaptured	12.5	6.3	
Rubery	Frequency in the natural population	63	528	591
nr Birmingham	Marked released	137	447	584
1953	Marked recaptured	18	140	158
	% of releases recaptured	13.1	31.3	
Rubery	Frequency in the natural population	53	486	539
nr Birmingham	Marked released	64	154	218
1955	Marked recaptured	16	82	98
	% of releases recaptured	25.0	52.3	

* Data from Kettlewell, H. B. D. (1955). Selection experiments on industrial melanism in the Lepidoptera. *Heredity, Lond.*, **9**, 323–42, and Kettlewell, H. B. D. (1956). Further selection experiments on industrial melanism in the Lepidoptera. *Heredity, Lond.*, **10**, 287–301.

Note that there are actually two melanic forms of this moth—*carbonaria* and *insularia*—but for the sake of this discussion the *insularia* form has been omitted from the data. The three forms of the moth are believed to be determined by three alleles of a gene, with *carbonaria* dominant to *insularia* and *typica*, and *insularia* dominant to *typica*.

we cannot distinguish between these explanations. When we also consider the Birmingham experiments, it is possible to discard the first three explanations because the results are almost exactly reversed. That is, a greater proportion of the melanic form is recaptured in the Birmingham experiments.

It is quite clear from these experiments that the form that is better camouflaged (in the particular habitat being studied) is about twice as likely to be recaptured as the form that is more obvious to us. Does a putative predator see the insects in the same way as we do? The answer is that they do, but it was to study this problem that Kettlewell chose to work with *B. betularia* in the first place. From the point of view of the technique that Kettlewell wanted to use, *B. betularia* has a very useful habit. It exposes itself during the day, resting on tree-trunks, and does not hide away in crevices or other dark places as do many other night-flying moths. This meant that Kettlewell was able to place moths of known phenotype in positions that they were quite likely to choose for themselves, whereas for the experimenter these positions were eminently suitable for observation from a 'hide'. Kettlewell placed equal numbers of typical and melanic females on tree trunks. He replaced those that were missing only when all of one phenotype had disappeared.

Using the laborious technique of direct observation through binoculars, Kettlewell noted that at Deanend Wood 190 specimens were taken by birds and, of these, 164 were melanic and 26 were typicals. At Rubery, on the other hand, the situation was reversed. Observation on the feeding by Redstarts (*Phoenicurus phoenicurus*) on equal numbers of the typical and melanic forms showed that the birds ate 43 typicals and only 15 melanics. As a result of this work, Kettlewell clearly established that the different forms of this moth were subject to visual selection by birds and that this selection was the likely explanation for why the melanic forms were common in industrial areas and rare in rural habitats. If this interpretation is correct it is reasonable to expect that the introduction of smokeless zones in the large cities would allow a reversal of the selective process, because the trunks of trees would no longer be as dark and the melanic form would be less cryptic. There is some evidence that the typical form *is* increasing in frequency in the Manchester area. During the period from 1952 to 1964 Michaelis collected 749 *B. betularia* at Didsbury and all were melanic. In the late 1960s Askew and colleagues trapped moths at Didsbury and the nearby district of Longsight, and among the 364 specimens they examined, eight were typicals. On its own this evidence is meagre, but a similar trend was observed by Cain and Sheppard at Caldy on the Wirral peninsula between 1959 and 1965, and this also coincided with the introduction of clean-air zones in that area. The work on industrial melanism in Manchester and Liverpool is continuing, and further evidence on the effect of the introduction of smokeless zones is eagerly awaited. On the other hand, some recent work shows that the frequency of *carbonaria* increased in southern England between 1952 and 1970!

9.7.2 The evolution of dominance

Other work by Kettlewell on *B. betularia* can be quoted to demonstrate the genetic principle called the evolution of dominance. Up to now you have probably assumed that the dominance relationship between a pair of alleles is inflexible and not subject to modification. In the 1920s Fisher became convinced that dominance was not a fixed property, but his ideas were subjected to severe criticism because it was (and indeed still is) difficult to explain in biochemical terms how changes in the dominance relationships between alleles can come about. On the other hand, there are now several examples of the evolution of dominance and one of the best is in *B. betularia*. This example also demonstrates that dominance is a property of characters and *not of the genes themselves*.

Among the collections of moths stored in various museums in England there are melanic forms of *B. betularia* that were captured about 1860. These specimens have more light-coloured scales than those found today and are therefore intermediate between the modern *carbonaria* and *typica* forms.

Many heterozygotes caught this century have four white specks on the head and at the base of the forewings. This was the evidence for Haldane's suggestion that the evolution of forms with entirely black scales must have been brought about by the selection of *modifying genes*; that is, the expression of melanism depended on the genetic background in which the melanic allele acted.

modifying genes

Kettlewell knew that Ford had modified the intensity of pigmentation in the dark forms (*curtisii*) of the lesser yellow-underwing moth (*Triphaena comes*) by selective breeding and this encouraged him to test Haldane's hypothesis by three elegant, but very time-consuming experiments. Two of these experiments involved the transfer of the melanic allele into an alien genetic background by the standard method of repeated back-crossing to the alien type. The third experiment involved transferring the melanic allele back again into its normal background. The outline of the experiments is given in Tables 10 and 11.

Table 10 Programme for changing the genetic background of the allele for melanism in *B. betularia*

1	Melanic (Birmingham) × typical (Cornwall) for three generations
2	Melanic (Birmingham) × typical of the Canadian species *Biston cognataria* for four generations
3a	Melanic from third generation in Experiment 2 × typical (Birmingham)
3b	Melanic from third generation in Experiment 2 × typical (Cornwall)

Table 11 The breeding programme of repeated back-crossing used by Kettlewell

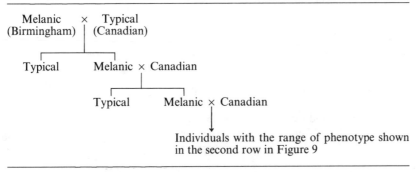

Individuals with the range of phenotype shown in the second row in Figure 9

QUESTION After three generations of back-crossing melanic individuals from Birmingham to *typica* from Cornwall, what proportion of the genotype of a melanic individual will be of Cornish origin?

ANSWER There are two distinct parts to this problem.

(a) Firstly we ignore the chromosome carrying the melanic allele. It is easy to see that after one generation of back-crossing the genotype of a melanic moth will be half Cornish. After the next cycle it will be three-quarters and after the third cycle seven-eighths Cornish.

(b) Because we are using melanic individuals for successive backcross generations we are necessarily selecting for the retention of the Birmingham chromosome carrying the melanic allele. The only way by which Cornish genes can be introduced into this chromosome is by crossing-over at meiosis.

It is clear, therefore, that Kettlewell's experiment does not entirely change the genetic background, but as you will see shortly even this limited adjustment of the genetic background is sufficient to change the phenotypic expression of the melanic allele in all three experiments. Using the segregating progeny of a cross between melanic and typical *B. betularia* from Birmingham as his reference, Kettlewell found that melanic moths from his first experiment (Table 10) were all jet black and *peppered with white dots*, whereas the Birmingham melanics were *totally black*. This suggests that the genetic make-up of the Cornish populations was different, but not all that different, from the Birmingham populations

Again, using the segregating progeny of a cross between melanic and typical *B. betularia* from Birmingham as his reference (row 1 in Fig. 9) Kettlewell found, in his second experiment (Table 10 and row 2 in Fig. 9), a range of phenotypes between the extremes of the modern melanic and typical forms from Birmingham. We must conclude, therefore, that in Britain there is a system of modifying genes which keeps the two forms of the moth clearly distinct, and that on the introduction of other genes this genetic complex is disrupted, so that it is no longer capable of maintaining the clear dominance relationship between the melanic and typical forms.

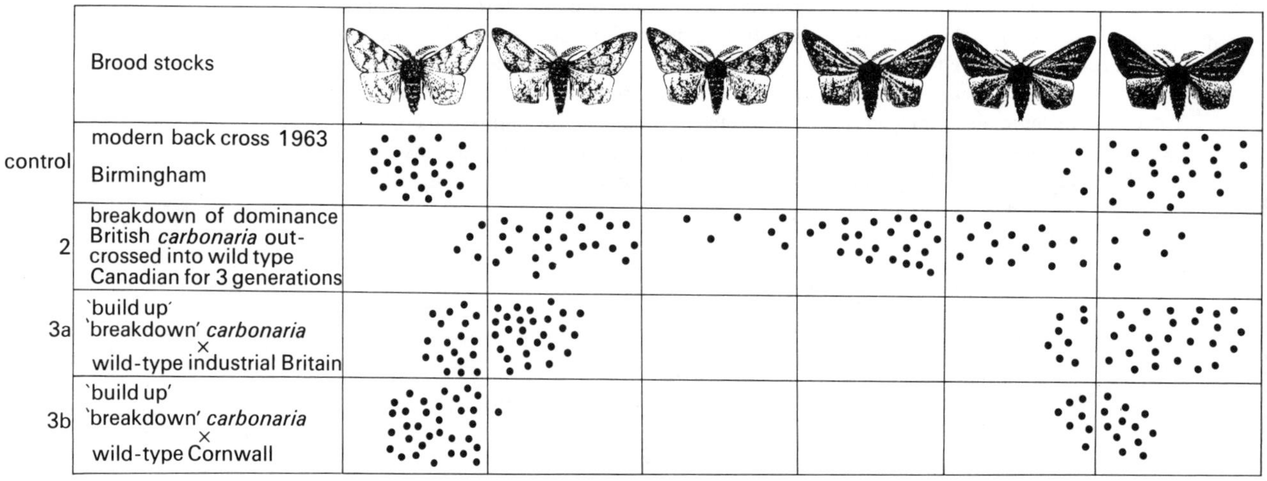

Figure 9 Distribution of phenotypes into six categories in broods showing 'breakdown' and 'build-up' of dominance. The position of each dot records the degree of expression of melanism in an individual moth. (Crippled and worn insects were not scored.)

Experiments 3a and 3b (Fig. 9) reintroduce the melanic allele into a British background.

QUESTION What conclusion can be drawn from the results of experiments 3a and 3b shown in Figure 9?

ANSWER The dominance relationship between the *carbonaria* and *typica* alleles is virtually restored to normal when only 50 per cent of the genetic background is British. Both the Birmingham and Cornish populations must contain a similar array of genes capable of modifying the dominance of the *carbonaria* allele.

We can draw the following conclusion. An allele has been passed from one genetic background to another and then back again. Each time, the phenotypic expression of the allele in terms of its dominance relationship with the alternative allele is changed. Dominance cannot, therefore, be a property of the gene itself but only of the gene complex in which it acts.

Summary of industrial melanism

So far in these Units we have spent rather longer on this one example than on any othe . It is by no means the only example of the various genetic and evolutionary phe.iomena we have described, but by concentrating on industrial melanism we have been able to demonstrate selection in action, protective colouration, the use of breeding tests for purposes other than for describing the mode of inheritance of a character, as well as the change of dominance relationship between alleles as a result of changing the genetic background.

9.7.3 DDT resistance in the yellow-fever mosquito

In Section 9.8 and in Unit 10 there will be several more examples of selective processes occurring in natural populations of animals and plants. But there are three further principles of population genetics that need to be explained first. Although we could use *B. betularia* again as the example, a study of both economic and scientific importance with the yellow-fever mosquito, *Aedes aegypti*, makes the points more forcibly.

In 1956 a strain of the mosquito was collected on Penang Island, Malaya, from a district in which insecticides had not been applied. It was found that the larvae had a natural tolerance to DDT (LC_{50} of 0.08 p.p.m.)* significantly in excess of strains from other parts of the world where DDT had been used. This is surprising because in contrast with many strains of this species, which had gained little or no DDT resistance during experiments in the laboratory, this Penang strain soon gained DDT resistance in initial generations of selection. It did, however, revert to susceptibility equally rapidly on relaxation of the selection pressure.

Abedi and Brown were concerned that the phenomenon could not, at first sight, be explained by normal selective processes and so they studied the problem further. During their experiments they *sibmated* (that is, mated brothers and sisters) their mosquitoes and subjected entire generations of larvae either to selection by DDT or to no selection. The selection was imposed by subjecting larvae to concentrations of DDT that would kill between 70 per cent and 90 per cent of them. The survivors were allowed to mature and their progeny were tested for susceptibility to DDT. At the same time, however, they counted the number of eggs laid, the number of larvae hatched and the number of mature larvae produced by each female mosquito, and these data gave a clue to the solution of their problem. They found that the female mosquitoes that gave rise to the most resistant progeny produced fewer eggs than females whose progeny were less resistant (see Table 12). This suggests that the females that survived selection by DDT when they were larvae had a lower reproductive fitness than the susceptible females *when the selection was relaxed*. That is, there was an association between DDT resistance and fecundity. It is not possible to say whether the two characters were genetically linked, but fortunately Abedi and Brown continued their experiments (see Table 13) and found that by the ninth generation the rapid reversion to susceptibility on relaxation of selection had disappeared.

QUESTION How would you account for the apparent loss of the ability to revert to susceptibility?

ANSWER It is likely that by this stage of the experiment reproductive fitness had become separated from DDT resistance.

Estimates of the reproductive capabilities of the DDT-resistant females of the G_9 generation proved that these mosquitoes actually produced more progeny than their DDT-susceptible relations! Further breeding work showed that the DDT resistance depended primarily on an allele of a single gene and that successive stages in the selection programme involved evolution of dominance in the direction of greater resistance.

* Normally the toxic effects of poisons are measured as LD_{50} (lethal *dose* at which 50 per cent of the individuals die); here we can be much more precise and quote a lethal concentration (LC_{50}) because the natural habitat of the larvae is water.

Table 12* DDT tolerance levels of successive generations of the Penang strain of *A. aegypti*

Generations	LC_{50}/p.p.m. (on larvae)
Parental	0.08
G_1 selected	0.12
G_2 selected	3.5
G_3 relaxed	0.13
G_4 relaxed†	0.09
G_4 selected‡	3.6
G_5 relaxed	3.8
G_6 selected	1.19
G_7 relaxed	0.47
G_8 selected	0.52

* Data from Abedi, Z. H., and Brown, A. W. A. (1960). Development and reversion of DDT-resistance in *Aedes aegypti. Can. J. Genet. Cytol.*, **2**, 252–61.

† 'G_4 relaxed' means that no selection was applied to the preceding generation.

‡ 'G_4 selected' means that selection was applied to the larvae of the preceding (G_3) generation, whereas the LC_{50} test was on the larvae of the G_4 generation.

sibmating

Table 13* DDT tolerance levels of successive generations of the Penang strain of *A. aegypti* (Table 12 continued)

Generations	LC_{50}/p.p.m. (on larvae)
G_8 selected	0.52
G_8 relaxed	0.26
G_9 selected	18.00
G_{10} relaxed	18.3
G_{10} selected	28.0

* For source details, see Table 12.

Summary

The example of DDT resistance in *A. aegypti* demonstrates three principles of population genetics.

1   Populations of organisms often contain mutant forms that are of no immediate use in the relatively stable conditions of the natural habitat, but they may be vital to the survival of the population if the environment, and consequently the selection pressure, changes markedly.

2   The individuals that possess, fortuitously, characters that enable them to survive under certain conditions of selection are not necessarily the most fecund of the population.

3   The response to selection depends not only on individuals possessing suitable genes (in this case resistance to DDT), but also on having these and other genes in the right combinations (associated with genes concerned with the fitness characters of egg number, hatchability and larval survival, etc.).

A note of warning: the occurrence of some mutant forms that are normally of no obvious use unless there is a change in selection is often called *preadaptation*. We have deliberately avoided using this term because it is very easy to misinterpret its meaning. It implies that the organism *anticipated* the selective change and was *prepared* for it. Clearly this is not true.

preadaptation

You should now attempt SAQs 4 and 5.

9.8   The statics of evolution

At the end of Section 9.6.3 we stated that further discussion of genetic polymorphism would be postponed until after we had considered some examples of evolution in action. The time has now come to consider genetic polymorphism in more detail.

> **ITQ 5**   How is it possible to maintain a pair of alleles in a population at a frequency above mutation rate?

We have seen that not only is selection a major agent causing evolutionary change, but it is also involved in maintaining a genetic polymorphism when there is no net change in allele frequency (see special case (d), p. 423). The evolutionary process appears to be both stable and static. Are there other ways by which a genetic polymorphism can be maintained?

9.8.1   What maintains a genetic polymorphism?

We have seen from our generalized model for selection that a stable equilibrium can exist for a pair of alleles at a locus when the heterozygote is placed at an advantage relative to the two homozygotes. On the other hand, sexual dimorphism in humans is a stable situation and yet there is only one homozygote: the YY-chromosome type is absent. Obviously our simple algebraic model of a balance of selective forces is inadequate to explain all situations, but an extremely useful and all-embracing verbal definition of a genetic polymorphism was devised by Ford in 1940. He regards a *genetic polymorphism* as 'the occurrence together in the same interbreeding population of two or more forms of the same species in such proportions that the rarest of them cannot be maintained merely by recurrent mutation'. This definition has withstood the test of time even though we now know of many different processes by which a genetic polymorphism can be maintained.

genetic polymorphism

> QUESTION   Using only examples that have been referred to elsewhere in these Units, make a list of those variation patterns that are likely to be genetic polymorphisms.
>
> ANSWER   (a) HbA and HbS haemoglobins in humans (p. 404);
> (b) eye colour in humans (p. 399);
> (c) sexual dimorphism in mammals (p. 398);
> (d) hair shape in humans (straight, wavy or curly) (p. 399);
> (e) the spot/non-spot situation in the butterflies in ITQs 2 and 4 (p. 409);
> (f) industrial melanism in *B. betularia* (p. 426).
> Other examples are illustrated in Figure 6(iii), facing p. 419.

When a genetic polymorphism is effectively at equilibrium it is called a *balanced polymorphism*. On the other hand, the polymorphism must have evolved from an original monomorphic situation and so there would have had to have been a transient stage between monomorphy and the balanced polymorphism, that is a *transient polymorphism*. Whether any genetic polymorphism is truly balanced is a matter for debate, but it is reasonable to argue that, in industrial areas, the polymorphism of melanism in *B. betularia* is a balanced polymorphism now, whereas it was a transient polymorphism during the latter part of the last century and also at the beginning of this one. As we saw on p. 428, there is some evidence that the frequency of the *carbonaria* form is declining in the Manchester and Liverpool areas, and so it appears that the polymorphism is now transient in the reverse direction.

Our model has assumed constant selective values and yet it is naïve to expect selection not to vary. When a region is broken up into differing environments that merge with each other in a graded way, it is likely that there will be a corresponding gradation in the ability of individuals to exploit these habitats. Where the frequency of a particular phenotype is related to some habitat factor that varies in this way, there is a gradual change of frequency related to distance (that is, a cline). We presume that the selective pressure varies from one end to the other, whereas at any particular point in the cline the selection may be relatively constant.

balanced polymorphism

transient polymorphism

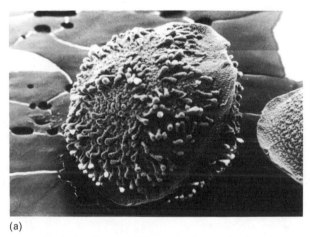
(a)

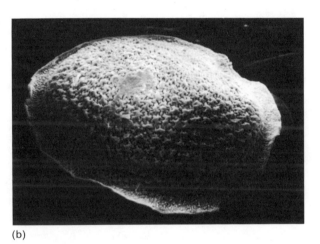
(b)

Figure 10 (a) Papillate and (b) non-papillate seeds of corn spurrey, *Spergula arvensis*. (Photograph taken with a scanning electron microscope; magnification: (a) ×80; (b) ×72.

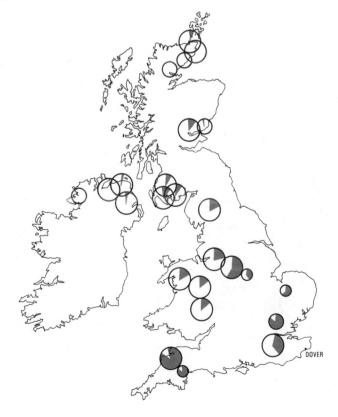

Figure 11 Map showing the proportions of papillate- and non-papillate-seeded plants of *Spergula arvensis* in different localities in Britain.

White sectors: proportions of non-papillate seeds; black sectors: proportions of papillate seeds; large circles: 100 plants or more; medium circles: 50 to 100 plants; small circles: 10 to 50 plants.

Let us explore this further. New studied the distribution and the genetics of the seed-coat character 'papillate' in corn spurrey *Spergula arvensis* (see Fig. 10). This is an annual herb, which reproduces only sexually. She showed that a pair of alleles was involved and that the heterozygous plants produced seeds with an intermediate number of papillae. Figure 11* shows the distribution of papillate-seeded plants (the homozygotes and heterozygotes have been grouped together) in Britain in 1955 and 1956. Clearly the frequency of the papillate form declines from south to north. Figure 12 demonstrates even more impressively that the cline is related to latitude.

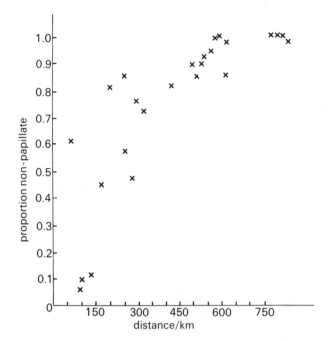

Figure 12 Graph showing the correlation between the proportion of non-papillate-seeded plants and the distance NNW from Dover (SE England).

One of the more obvious climatic variables in Britain that is related to latitude is temperature, and Figure 13 is a map showing the average July daily temperature for the thirty-year period 1931–1960. New found that there was a relationship

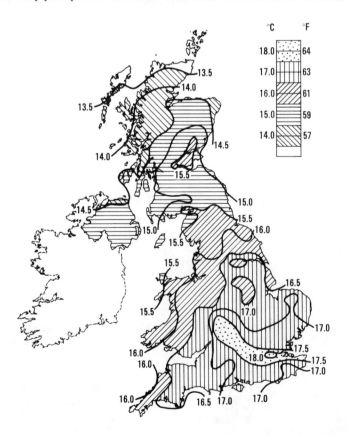

Figure 13 The mean July daily temperature in the United Kingdom, 1931 to 1960.

* Figures 11, 12 and 14 are based on illustrations from New, J. (1958). A population study of *Spergula arvensis. Ann. Bot.*, **22**, 457–77.

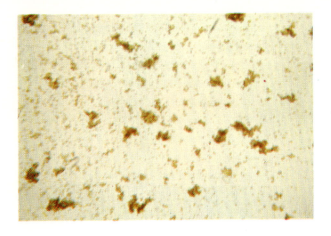

Figure 2   Agglutination of human red blood cells.

Figure 16 Mimetic and non-mimetic butterflies and their models.

(1) *Danaus chrysippus*, the model for 3.
(2) *Amauris niavius*, the model for 4.
(3) *Papilio dardanus* (race *cenea*) ♀, form *trophonius*.
(4) *Papilio dardanus* (race *tibullus*) ♀, form *hippocöonides*.
All these four butterflies are sympatric.

(5) *Papilio dardanus* (race *meriones*) ♀, non-mimetic form.
(6) *Papilio dardanus* ♂, always non-mimetic.
(7) *Limenitis archippus archippus*, mimic of 8.
(8) *Danaus plexippus plexippus*, model for 7.
Note that although we have used the term race here and in the text, some authorities on these butterflies prefer to use the term sub-species.

between the frequency of plants producing non-papillate seeds and the July temperatures (Fig. 14), and this prompted her to test the effects of temperature on fitness characters in the plant.

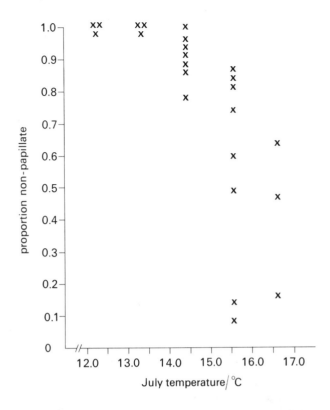

Figure 14 Graph showing the correlation between the proportion of non-papillate-seeded plants and the July temperature.

Experiments on germination and survival showed that non-papillate seeds germinate more rapidly than papillate seeds at 13°C, whereas the reverse is true at 21°C. Non-papillate-seeded plants produce a lower proportion of fertile capsules than papillate-seeded plants when grown at 21°C. Non-papillate-seeded plants wilt and die at the higher temperature and low humidity (55 per cent), whereas papillate-seeded plants grow normally under these conditions.

QUESTION Suggest a hypothesis to explain the cline.

ANSWER At the southern end of the country the plants that produce non-papillate seeds are at a disadvantage both vegetatively and reproductively when the weather is warm. At the lower temperature expected at the northern end of the cline, the non-papillate seeds germinate more rapidly and, presumably, become established as seedlings and young plants to the exclusion of the later germinating papillate seeds.

It is very likely that the change in frequency of the papillate-seeded form of *S. arvensis* from south to north Britain depends on summer temperature. Temperature does, of course, fluctuate violently both diurnally and seasonally, but if we take the temperatures used by New as a guide, they show that a range of 13°–21°C is sufficient to invert the selective advantage of the non-papillate and papillate forms.

Summary

It should now be clear that (a) the relative fitness of the two morphs changes in parallel with changes in mean temperature and (b) we can measure the selection involved in maintaining the polymorphism at any one place quite independently of the rest of the cline.

10.1 Genetic polymorphism

Having considered in some detail melanism in *B. betularia* and papillate seeds in *S. arvensis*, and discussed some of the suggestions made about how the polymorphisms are maintained, we can now ask two questions. How widespread is genetic polymorphism? Is it important in evolution?

Prior to 1966, estimates of the extent of polymorphic variation in animals and plants were related primarily to morphological variants, with biochemical variants like the blood groups and the haemoglobins being the exception rather than the rule. It was Hubby and Lewontin who in 1966 first showed that *biochemical polymorphism* was likely to be very widespread, when they demonstrated that variation in enzymes and other proteins in *Drosophila pseudoobscura* was produced by at least 8 of the 21 loci they examined. They used the technique of acrylamide-gel electrophoresis. Population analysis suggested that, when the loci to be examined are chosen at random, the average population is polymorphic for about 30 per cent of all loci!

biochemical polymorphism

Since the publication of this work many people have studied allozyme* polymorphisms in many different organisms. In consequence, we now know that this kind of polymorphism is very common and widespread and that numerous organisms are heterozygous at many loci. Current work is attempting to answer the question: why is there so much heterozygosity at these various enzyme loci?

Already we have posed three problems in this Section. We shall begin to examine them by considering an example of chemical polymorphism in a fungus.

10.1.1 Tyrosinase in *Neurospora*

It is unfortunate that during the 1950s and early 1960s population and ecological geneticists were not particularly interested in fungi, because work by Horowitz and his colleagues with *N. crassa* indicated that allozyme variation might be common and was reflected in characteristic differences in enzyme activity or stability. Indeed Horowitz was studying two forms of tyrosinase that had different thermal stabilities (one had a half-life of 5 minutes at 59°C, whereas the other had a half-life of 70 minutes at the same temperature). It was only later that he found that the two forms differed in their electrophoretic properties at pH 6, the stable form migrating towards the cathode more slowly than the unstable form.

Of itself, this discovery is of no great interest to the population geneticist, but fortunately Horowitz was intrigued by this variation and he wondered whether the enzyme might be highly variable in natural strains of *N. crassa*. He collected a dozen strains from the wild and screened the tyrosinase they produced with respect to thermostability and electrophoretic migration on paper. In addition to the two forms he had described previously he found two more (Table 14). Horowitz crossed

Table 14*   Properties of the four tyrosinases of *N. crassa*

Type	Thermostability/min (half-life at 59°C)	Electrophoretic migration/m $h^{-1} \times 10^{-3}$ (on paper at pH 6)
S	70	2.00
L	5	2.25
PR-15	20	1.50
Sing-2	70	1.50

* Data from Horowitz, H. H., Fling, M., Macleod, H., and Watanabe, Y. (1961). Structural and regulative details controlling tyrosinase synthesis in *Neurospora*. *Cold Spring Harb. Symp. quant. Biol.*, **26**, 233–8.

S and L are the original variants that were discovered in descendants of isolates of opposite mating type obtained by Abbott in Louisiana. Horowitz obtained PR-15 and Sing-2 as wild isolates from Puerto Rico and Singapore.

* The term *allozyme* will be used in these Units for situations where enzymes with different electrophoretic properties are produced by alleles of the same gene. *Isozymes* is too loose a term in the context of these Units, because it includes both similar enzymes produced by different loci and different forms of one enzyme with the same basic function whose genetic determination has yet to be worked out.

allozyme

the various strains and showed that the four tyrosinases appeared to be produced by alleles of a single gene. Heterokaryons (see *Life Cycles**) between two strains produce a mixture of both types of enzyme. He concluded† 'It is tempting to ascribe this polymorphism to the weakness of the selective forces acting on a non-essential enzyme‡, but in the absence of comparable studies on any essential enzyme of *Neurospora*, this must remain conjecture'. Other people have subsequently worked with essential enzymes, in *Neurospora* and in many other organisms, but we must leave discussion of this work until Section 10.6.

Horowitz showed that tyrosinase production was controlled by a multiple-allele system. Do multiple alleles complicate the population models for the diploids that we have been considering, and particularly the Hardy–Weinberg equilibrium?

10.1.2 Multiple alleles in *Cepaea* [B]

So far we have concentrated our attention on loci that are represented in a population by not more than two alleles. Your knowledge of the origins and nature of changes, or mutations, in the genetic material must convince you that there is no reason why further mutations to other alleles should not arise at such a diallelic locus. Of course, no more than two different alleles may be present in any single diploid individual, but more than two different alleles may be represented at a particular locus among a population of individuals.

In Section 9.4.2 we showed that if two alleles, A_1 and A_2, are present in a random-mating population at frequencies of p and q, then the genotype frequencies will be $p^2 A_1 A_1$, $2pq A_1 A_2$ and $q^2 A_2 A_2$. These genotype frequencies remain constant in the absence of mutation, migration, selection and random processes. Is such a Hardy–Weinberg equilibrium state possible for a locus with more than two alleles? The answer is 'yes'.

For the simple case of three alleles A_1, A_2 and A_3 with frequencies p, q and r (where $p + q + r = 1$) the array of possible genotypes will have frequencies at Hardy–Weinberg equilibrium given by the algebraic expansion of $(p + q + r)^2$. This is a simple extension of the result we obtained for the two-allele case.

Now

$$(p + q + r)^2 = (p + q + r)(p + q + r)$$
$$= p^2 + q^2 + r^2 + 2pq + 2pr + 2qr$$

After one generation of random mating, therefore, the three homozygotes $A_1 A_1$, $A_2 A_2$ and $A_3 A_3$ will be present with frequencies p^2, q^2 and r^2, respectively, and the frequencies of the three heterozygotes $A_1 A_2$, $A_1 A_3$ and $A_2 A_3$ will be $2pq$, $2pr$ and $2qr$, respectively. These equilibrium frequencies will be maintained as long as mating is random and as long as the allele frequencies are not perturbed by selection, mutation, migration or random processes. The general case can now be stated simply. If we have r alleles $A_1, A_2, \ldots, A_r$ at a single locus with frequencies $p_1, p_2, \ldots, p_r$, respectively, then at equilibrium the frequency of any homozygote $A_i A_i$ will be p_i^2 and the frequency of any heterozygote $A_i A_j$ (where $i \neq j$) will be $2p_i p_j$.

Shell colour in the snail *Cepaea nemoralis* is a polymorphism determined by three alleles at a single locus. C^B (brown) is dominant to C^P (pink) and both are dominant to C^Y (yellow) (see broadcast notes to Television programme 9).

QUESTION Is it possible to determine the genotype of any individual *C. nemoralis* merely by looking at its phenotype?

ANSWER Because this series of three alleles displays a full hierarchy of dominance (that is, C^B dominant to C^P, and *both* dominant to C^Y), every heterozygote will resemble a homozygote in its phenotype, as indicated in the following Table.

---

* The Open University (1976) S299 LC *Life Cycles*, The Open University Press. This folder, which contains details of organisms mentioned in the Course, is part of the supplementary material for the Course.

† See Table 14 for source.

‡ Tyrosinase is an essential enzyme only during the development and maturation of the female fruiting bodies or when the fungus is grown under conditions of semi-starvation.

Phenotype	Possible genotypes		
brown	$C^B C^B$	$C^B C^P$	$C^B C^Y$
pink	$C^P C^P$	$C^P C^Y$	
yellow	$C^Y C^Y$		

This means that it is not possible, without embarking on a breeding programme, to ascribe a genotype to each snail, except for those with yellow shells, which must be homozygous $C^Y C^Y$.

ITQ 6 If the frequencies of alleles C^B, C^P and C^Y in a particular population are respectively p, q and r, what are the expected frequencies of brown-, pink- and yellow-shelled snails if the population is in Hardy–Weinberg equilibrium?

Suppose we visited a natural population of *C. nemoralis* and collected a sample of 100 snails comprising 19 browns, 45 pinks and 36 yellows, how would we estimate the frequency of the three alleles C^B, C^P and C^Y in this population? *If* we assume that the population was in Hardy–Weinberg equilibrium, we could, with the aid of some simple algebra, derive expressions relating the allele frequencies to the observed phenotypic frequencies. Let the frequencies of brown, pink and yellow alleles be p, q and r, respectively, and the proportions of brown, pink and yellow snails in the sample be a, b and c. We can construct a table as follows.

| | Phenotype | | | |
	Brown	Pink	Yellow	Total
Observed proportion	a	b	c	1
Expected frequency	$p^2 + 2pq + 2pr$	$q^2 + 2qr$	r^2	1

We wish to express p, q and r in terms of a, b and c. We have three independent equations relating three unknowns

$$p^2 + 2pq + 2pr = a$$
$$q^2 + 2qr = b$$
$$r^2 = c$$

so we may solve for p, q and r.

All yellow-shelled snails are $C^Y C^Y$ homozygotes and the yellow allele frequency may be estimated immediately as

$$r = \sqrt{c}$$

The pink allele frequency is not so straightforward because there is no distinguishable phenotypic class composed solely of $C^P C^P$ homozygotes (some of the pink-shelled snails will be $C^P C^Y$ heterozygotes). We know, however, that

$$q^2 + 2qr = b$$

Adding r^2 to each side produces a more manageable equation. Thus

$$q^2 + 2qr + r^2 = b + r^2$$
$$(q + r)^2 = b + r^2$$
$$q + r = \sqrt{(b + r^2)}$$
$$q = \sqrt{(b + r^2)} - r$$

But $r = \sqrt{c}$. Therefore

$$q = \sqrt{(b + c)} - \sqrt{c}$$

Finally, since the allele frequencies must sum to unity, the C^B frequency may be found by subtraction; that is

$$p = 1 - q - r$$
$$= 1 - (q + r) = 1 - [\sqrt{(b + c)} - \sqrt{c} + \sqrt{c}]$$

Therefore

$$p = 1 - \sqrt{(b + c)}$$

ITQ 7 Calculate the frequency of the C^B, C^P and C^Y alleles in our sample of 19 brown, 45 pink and 36 yellow-shelled snails.

Notice that in order to estimate allele frequencies we have had to assume that the population is in Hardy–Weinberg equilibrium. It is not possible, therefore, to use these estimates to test whether or not the population really is in Hardy–Weinberg equilibrium.

Not only do the shells of *C. nemoralis* vary in colour: they also show a variety of banding patterns. But rather than continue the discussion of multiple alleles in *Cepaea*, which is discussed at length in Television programme 9 and in the associated notes, we shall go on to consider other examples of genetic polymorphism.

You should now attempt SAQs 6 and 7.

10.1.3 Batesian mimicry

One of the most intriguing phenomena in nature is that of mimicry. In *Batesian mimicry* there are a number of distinct warningly coloured species, distasteful to predators or protected in some other way, acting as *models* which are mimicked by the polymorphic forms of a single more edible species. *Müllerian mimicry*, on the other hand, depends on a number of distasteful species, all of which are protected because they look like each other. We shall concentrate on Batesian mimicry in what follows, but some principles common to most kinds of mimicry will also be mentioned. Mimicry is an exercise in biological bluffing—rather like clothes on humans—or you'll never get to Heaven in powder and paint, 'cos the Lord don't like you as you ain't! Although widespread, it is probably most common between insect species, particularly butterflies and moths. The insects are predated by birds and it is clear that the mimicry has evolved in response to selective pressure by birds. Much of the early work on mimicry relied to a certain extent on a *belief* that Batesian mimicry works to the advantage of the mimic. Fortunately, there is experimental evidence from work using both genuine and artificial mimics and models. An elegantly simple experiment by van Zandt Brower with the butterflies *Limenitis archippus archippus* (the viceroy) and *Danaus plexippus* (the monarch) and the Florida Scrub Jay shows that mimicry works. In a series of experiments she offered butterflies in sequence to the birds starting with *L. archippus archippus* (mimic). The birds ate them. Each bird tasted the first *Danaus plexippus* (model) offered, spat it out and refused to eat any others offered. After offering *D. plexippus* 50 times in sequence she then substituted *L. archippus archippus*. The birds now avoided the tasty mimic completely. These butterflies are illustrated in Figure 16, facing p. 435.

An important bonus of this work was that van Zandt Brower observed that the individual birds vary in their ability to distinguish models from mimics. This suggested that the experiments should be repeated using potential mimics. *Limenitis archippus floridensis* is a butterfly that resembles *D. plexippus* but not very well. (It actually mimics another *Danaus* species.) She found, however, that the resemblance between *L. archippus floridensis* and *D. plexippus* was sufficient to deter the birds that had been through the previous experiment. Thus birds can mistake a poor mimic for a model belonging to a related mimetic complex, indicating that there is a selective advantage even of incipient mimicry. It had been maintained previously that, to be effective, mimicry had to be perfect, because it was so difficult to see how it could have developed or evolved. van Zandt Brower had now shown experimentally, however, that mimicry does not have to evolve in a single step; it can be built up and perfected over many generations. This result has been confirmed by several other studies using partial mimics; indeed there is good evidence to suggest that even the slightest mimic is at an advantage over a non-mimetic form *under laboratory conditions*.

QUESTION The minimum number of species involved in a mimetic complex is two, the model and the mimic, and these species can occur at varying densities, not necessarily dependent on each other. Is mimicry likely to work if the mimic is commoner than the model?

ANSWER Fisher suggested that it would not. He based his hypothesis on the argument that if the mimic is too common a bird will not have the opportunity to meet the model often enough for it to learn that an insect with a particular pattern may be distasteful.

Batesian mimicry

model
Müllerian mimicry

439

van Zandt Brower tested this hypothesis using mealworms *Tenebrio mollitor* and European starlings *Sturnus vulgaris*. She painted mealworms green or orange. Some of the green ones were dipped in 60 per cent aqueous quinine dihydrochloride and others in distilled water. The orange ones were dipped in distilled water only. Ten times per day, over a period of 16 days the birds were offered one green and one orange mealworm. Starling number 1 received nine distasteful green mealworms per day whereas Starlings 2, 3 and 4 received respectively 7, 4 and 1 distasteful mealworms. The four Starlings used ate all of the non-mimetic edibles (orange) in all 160 trials. Other food was available throughout the experiment and so the birds were not hungry. After initial tasting and violent rejection, the models were generally recognized by their appearance and avoided, and in consequence their mimics proved to be protected also.

The results of one series of trials are given in Table 15.

Table 15* Data for the reactions of Starlings to mimics

	Bird 1	Bird 2	Bird 3	Bird 4†
Mimics	(10%)	(30%)	(60%)	(90%)
Eaten	1	10	18	75
Killed	0	1	0	1
Pecked	0	10	6	4
Not touched	15	28	72	10
Totals	16	49	96	90
Percentage not eaten or killed	94	77.5	73	15.5

* Data from van Zandt Brower, J. (1960). Experimental studies of mimicry. IV The reactions of Starlings to different proportions of models and mimics. *Am. Nat.*, **94**, 271–82.

† Bird 4 was used for only 100 trials in ten days.

QUESTION While admitting that the mimic is a 'perfect' mimic in these experiments, what conclusions can be drawn from the data?

ANSWER (a) No mimic receives 100 per cent protection.

(b) The evidence that birds peck or kill mimics implies that they still make 'mistakes'. Clearly this enables a bird to respond to changes in the relative proportions of models and mimics.

(c) The proportion of mimics not killed or eaten gives an estimate of the per cent effectiveness of the mimicry. Plotting this against the percentage of mimics gives the graph shown in Figure 15.

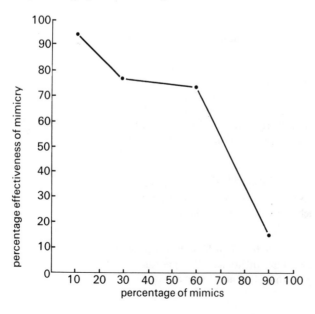

Figure 15 The effectiveness of mimicry for different proportions of models and mimics. The points represent the proportions of mimics not eaten or killed.

440

From Figure 15 we can estimate that when mimics comprise 50 per cent of the green mealworms offered the mimicry is 80 per cent efficient; indeed even with 70 per cent mimics the mimicry is 50 per cent effective. Because the mimic is perfect, these figures must represent the upper limit to the effectiveness of the mimicry. We can conclude, therefore, that when the model is sufficiently distasteful, it may give protection even when considerably outnumbered by its mimics, provided that their resemblance to it is an accurate one.

Evolution of mimicry

Whichever type of mimicry one studies, it is found that the resemblances between model and mimic may be extremely accurate and often involve a number of diverse characters such as colour, pattern, shape and even behaviour. Paradoxically these various phases are frequently controlled by a single switch-like mechanism. There are many genes involved but they are all inherited as a single unit! Furthermore, it is inconceivable that the complex mimetic patterns involved can have appeared coincidentally; the chances against such an event would be immense even if we were dealing with a single instance.

There have been three suggestions to explain this impasse. Punnett argued that corresponding genes in model and mimic have the same alleles. This is unlikely because mimicry is a phenotypic resemblance, and certainly there is evidence that deception is not achieved by pigments of the same chemical structure. Goldschmidt suggested that there are only limited ways that colour patterns can be developed and that mimicry is confined to nearly related members of a few systematic groups.

QUESTION From your general knowledge of biology give some examples that show that Goldschmidt's suggestion is nonsense.

ANSWER There are many, of which the following are a few:
beetles mimic wasps;
moths and butterflies mimic bees;
orchids mimic flies;
plant bugs (Hemiptera) mimic ants or moths;
caterpillars mimic snakes.

Earlier in this Section it was stated that van Zandt Brower showed experimentally that mimicry does not have to evolve in a single step. This had been predicted previously by Fisher and by Ford, and indeed was the basis of their theoretical model of the evolution of mimicry. Apart from the experimental evidence, genetic evidence obtained by breeding mimetic butterflies also shows that mimetic patterns do not arise in one step.

Some of the best examples occur in the butterfly genus *Papilio*—the swallowtails. The males of *Papilio dardanus* are invariably tailed, monomorphic and non-mimetic. The species is restricted to Africa, where it exists in eight or nine principal races, some of which overlap in their distribution. Thus

race *meriones* on Madagascar has females that look like males
race *antinorii* in Abyssinia has females that have tails but are otherwise unlike the males.

In all the other races the females do not have tails, some are mimetic and others are not, and the mimicry is effective. In some races the females are polymorphic, mimicking several different models.

By crossing the various allopatric* races, Clarke and Sheppard showed that the patterns of mimicry break down, producing intermediate forms that do not occur in the wild. For example, crosses between the form *trophonius* (see Fig. 16, facing p. 435) from South Africa with *hippocoön* from Ghana give females intermediate between *trophonius* and *hippocoön*.

It would appear, therefore, that balanced gene complexes have arisen in different parts of Africa and that the mimetic patterns are coordinated with a particular genetic background. The breeding work also shows that the various mimetic and non-mimetic female forms are controlled by closely linked genes whose effect is controlled by the sex of the individual. These closely linked gene complexes are

* *Allopatric/sympatric*. Sympatric races/species overlap in their distribution. Allopatric races/species do not, or very rarely, overlap.

allopatric   sympatric

known as *supergenes*. The F_2 progeny of the inter-racial crosses show no clear-cut segregations of the parental patterns. It is clear, therefore, that the mimetic pattern can only be produced when the 'right' genes are present. Clarke and Sheppard (see Sheppard, P.M. (4th edn 1975) *Natural Selection and Heredity*, Hutchinson, pp. 193–4) suggest that with *P. dardanus*, and also with *Papilio polytes* and *Papilio memnon*, the 'Batesian mimicry evolved as the result of the appearance of a new mutant and the subsequent improvement of the mimicry by the selection of modifiers'. In other words, the mimicry evolved by a series of steps, and this may suggest to you evolution of dominance and all the machinery for supergene formation. What evidence is there for this process?

We have already stated that the patterns are determined by supergene complexes. It so happens that one prominent feature of this mimicry is that where there is more than one mimetic female in a race there is complete dominance between the female forms. There are no intermediates. A similar situation occurs when sympatric races are crossed. For example, crosses between the form *trophonius* and *hippocoönides* (see Fig. 16, facing p. 435, for photographs of these forms) from South Africa give females that have the appearance of *trophonius*. Similar results were obtained for 14 out of the 21 morphs identified. The dominance relationship does not occur with the other seven morphs, which are (a) very rare, (b) mimicking rare models or (c) heterozygous and resemble some other polymorphic form but neither of the appropriate homozygotes.

Hybrids between allopatric races show that, like colour patterns, the absence of tails is incompletely dominant over the tailed condition. This is intriguing, because it shows that selection must be disruptive; that is, in some races there is selection for loss of tail in the females, thereby making it a better mimic, whereas selection favours, for some unknown reason, the retention of tails in males. Thus there is dominance for absence of tails in females, whereas in males there is dominance for presence of tails.

> **ITQ 8** How do the results of crossing sympatric and allopatric races differ for characters other than tails?

Summary of mimicry

In Batesian mimicry the model is relatively inedible or is protected in some other way; it is likely to have a conspicuous pattern, which acts as a warning to would-be predators. It must be common because otherwise a predator would have no chance to learn to avoid a beast with that particular pattern. Clearly the mimic and its model must be sympatric and the resemblance need extend only to visual characters like pattern and behaviour. The more closely the Batesian mimic resembles the model, the more likely it will be mistaken for the model. Therefore there will be selection favouring accentuation of the resemblance.

Batesian mimicry depends on predators with colour vision and reasonable memories hunting for food by sight.

10.1.4 Chromosome polymorphism in *Drosophila pseudoobscura*

Outside animal and plant breeding and problems of human genetics, the research geneticist is still in the fortunate position of being able to choose an appropriate organism to attack a particular problem. Rarely, however, has the taxonomist considered the genetic attributes of the species he has studied and so, not infrequently, the geneticist is faced with reclassification of the genera whose genetic systems he is studying. A particularly good example is the work of Dobzhansky, Sturtevant, and their colleagues with *Drosophila* species. Following on from the pioneering work of Morgan with *D. melanogaster*, they extended the study to wild populations of this fly genus. Almost immediately they were faced with unforeseen problems of identification and classification. While pursuing their genetic studies, they identified several new species and, of these, *Drosophila pseudoobscura* has proved to be one of the most interesting because of the widespread occurrence of inversions in chromosome III of this species. Dobzhansky spent much of his time during the past 40 years studying these inversions.

Refer back to Unit 5, Section 5.3 and refresh your memory about chromosomal inversions in Drosophila.

442

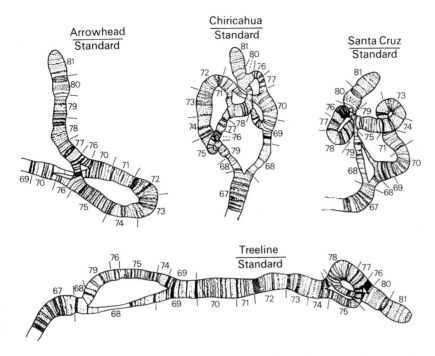

Figure 17 Drawings of the inversion heterozygotes of the chromosome III polymorphism as seen in the salivary gland cells of larvae of *Drosophila pseudoobscura*. The Standard arrangement in the relevant region runs from segment 67 to the terminus of the chromosome at segment 81.

Piñon Flats on Mount San Jacinto, California, is a desert slope, with piñon pine the dominant plant. Five gene arrangements of the third chromosome were found: Standard (ST), Arrowhead (AR), Chiricahua (pronounced Chirrikwah, CH), Tree Line (TL) and Santa Cruz (SC) (see Fig. 17). The names refer to the locations in which the particular chromosome arrangements were first found. The SC type accounted for only 0.5 per cent of the total and has been included with TL in Table 16. Note that there are six to eight generations a year in California, and that the larvae eat yeast. In all, 4 853 chromosomes were studied with over-all frequencies of 40.70 per cent ST, 25.14 per cent AR, 29.05 per cent CH and 5.11 per cent TL + SC.

QUESTION Each fly that is captured in the wild contains two copies of chromosome III. How would you use these captured flies to estimate the frequency of the inversion chromosomes in the population? Assume that all females captured will have mated.

ANSWER Provided the females are isolated as soon as they are captured and allowed to lay eggs, examination of any one of the larvae produced permits the determination of two homologous wild chromosomes.

Testing the males is a little more difficult. Here it is necessary to mate the males with females homozygous for a known chromosome III arrangement. Progeny larvae can then be scored for chromosome arrangement.

ITQ 9 How many larvae need to be scored so that there is only a 1/64 chance that a male heterozygote will be scored as a homozygote (see *STATS*, ST.12.1)?

Study the data in Table 16 for flies collected at Piñon Flats in California.

QUESTION What can you say about the relative frequencies of the chromosome arrangements throughout the year?

ANSWER They vary from month to month. To help you, the most frequent form has been set in italic. Apart from 1939 and 1941, the ST arrangement is commonest at the beginning and end of the year, but it gives way to the CH arrangement in late spring and summer. In 1939 AR was the commonest form in the summer, whereas in 1941 ST remained the commonest form throughout the year.

Analysis of χ^2 (for between months, within years) indicates that the probability that these fluctuations are the result of chance is less than 0.01.

There is also evidence that the frequency of the different chromosome arrangements changes with altitude. Thus further north in the Sierra Nevada of California the frequency of Standard decreases progressively from 46 per cent at an elevation of 260 m, to 10 per cent at an elevation of 3 000 m. Arrowhead changes from 25 per cent at 260 m to 50 per cent at 3 000 m, and Chiricahua fluctuates between 16 per cent and 20 per cent at the same altitudes (Fig. 18).

Table 16* **Percentage frequencies of gene arrangements Standard (ST), Arrowhead (AR), Chiricahua (CH) and Tree Line (TL), in the third chromosome of _D. pseudoobscura_ at Piñon Flats (elevation 1 200 m), California, USA**

Month/Year	ST	AR	CH	TL	n†	Month/Year	ST	AR	CH	TL	n†
4/39	50.8	29.5	13.1	6.6	61	3/41	56.2	10.9	23.6	9.1	110
5/39	27.9	35.8	30.0	6.3	240	4/41	58.2	20.0	17.3	4.5	110
6/39	29.9	35.1	30.5	4.5	154	6/41	32.9	32.3	32.3	2.6	192
8/39	35.9	33.3	25.6	5.1	156	8/41	51.9	21.3	25.9	0.9	108
9/39	51.1	22.6	23.2	3.2	190	10/41	56.2	18.8	16.3	8.7	80
10/39	54.6	25.4	16.6	3.5	284	12/41	45.0	24.0	24.0	7.0	100
3/40	44.6	20.2	30.1	5.2	386	4/42	51.0	21.6	19.6	7.8	102
4/40	34.7	28.4	33.5	3.4	176	5/42	48.0	17.0	25.0	10.0	100
5/40	28.2	27.2	39.6	5.0	202	6/42	29.8	22.8	40.4	7.0	114
6/40	24.1	30.0	41.8	4.1	170	7/42	41.9	21.8	30.6	5.6	124
9/40	34.6	25.0	37.5	2.9	104						
11/40	37.5	32.5	26.3	3.7	80	4/45	32.7	31.3	34.1	2.0	352
						3/46	56.5	18.3	18.3	7.0	558
						6/46	26.2	23.8	43.8	6.2	500

* Data from Dobzhansky, Th. (1947). A directional change in the genetic constitution of a natural population of _Drosophila pseudoobscura_. _Heredity_, _Lond._, **1**, 53–64.

† _n_ is the number of chromosomes examined.

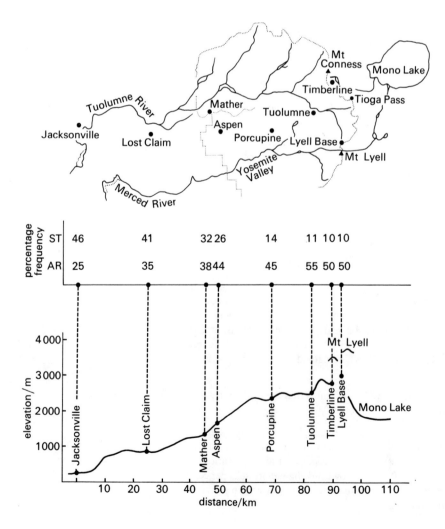

Figure 18 Frequency of Standard and Arrowhead inversion types of chromosome III of _D. pseudoobscura_ at different elevations in the Sierra Nevada (some 480 km north of Mount San Jacinto). The map above shows the position of the sites mentioned in the lower part of the Figure. The dotted line on the map is the boundary of Yosemite National Park, California.

Obviously we have a genetic polymorphism here. How is it maintained? Firstly, Wright and Dobzhansky have shown in population-cage experiments* that inversion heterozygotes are favoured at the expense of homozygotes. The data in Table 17 show that if one starts with 20 per cent homozygous AR and 80 per cent homozygous CH flies, then the relative frequencies of the two forms change until an oscillating frequency of AR is set up between 60 and 70 per cent. A situation of balanced polymorphism is obtained consistent with the relative fitnesses entered below the Table.

Table 17* Frequencies (per cent) of AR and CH third chromosomes in an experimental population of *D. pseudoobscura* kept at 25°C

Days	Generations	AR	CH	Days	Generations	AR	CH
0	0	20.0	80.0	365	14.6	67.7	32.3
35	1.4	40.7	59.3	488	19.5	69.0	31.0
70	2.8	41.3	58.7	649	26.0	66.3	33.7
105	4.2	55.3	44.7	858	34.3	60.7	39.3
270	10.8	67.7	32.3	1 240	49.6	62.0	38.0
				2 039	81.6	67.0	33.0

* Data from Beardmore, J. A., Dobzhansky, Th., and Pavlovsky, O. A. (1960). An attempt to compare the fitness of polymorphic and monomorphic experimental populations of *Drosophila pseudoobscura*. *Heredity, Lond.*, **14**, 19–33.

The flies were kept in population cages. Samples of 150 flies (300 chromosomes) were taken from time to time. The data are consistent with the following relative fitness of the three karyotypes.

AR/AR	AR/CH	CH/CH
0.71	1	0.43

When the biological fitness of monomorphic and polymorphic experimental populations are compared (Table 18), you can see that the polymorphic populations are more productive.

Table 18* Number of individuals and biomass in polymorphic and monomorphic experimental populations of *D. pseudoobscura*

		Polymorphic (AR + CH)	Monomorphic (CH)	Monomorphic (AR)
Number of	♀♀	184.3 ± 5.6	169.2 ± 7.5	139.1 ± 6.1
	♂♂	115.5 ± 5.5	101.8 ± 5.9	74.3 ± 4.5
Biomass/mg	♀♀	221.9 ± 6.1	204.1 ± 9.3	168.3 ± 6.9
	♂♂	103.4 ± 5.2	89.3 ± 5.8	62.8 ± 3.9

* Data from Dobzhansky, Th., and Pavlovsky, O. A. (1962). A further study of fitness of chromosomally polymorphic and monomorphic populations of *Drosophila pseudoobscura*. *Heredity, Lond.*, **16**, 169–77.

Similar experiments with CH and ST show heterozygous advantage when carried out at 25°C or 21°C, using *Kloeckera* yeasts. If, however, *Zygosaccharomyces* yeasts are used at 21°C, the heterozygous advantage vanishes. It is obvious, therefore, that the chromosome arrangements confer advantages on individuals only under certain conditions and, if these conditions change, it is advantageous for the population to be polymorphic. We must conclude that these physiological advantages override the lowered fertility of an inversion heterozygote (see Unit 5, Section 5.3.1).

It is not possible to predict the stabilization frequencies in experimental populations. For example the mixtures *ST and CH at 25°C stabilize at ST 60–70 per cent, CH 30–40 per cent*, but *ST and AR at 25°C also stabilize at ST 60–70 per cent, AR 30–40 per cent*.

* *Population cages* are specially designed boxes in which several thousand flies live and reproduce. Inserted into holes in the base of the box are small culture tubes containing *Drosophila* medium laced with yeast. These tubes can be removed from time to time without opening the box. Some can be used to sustain the colony, whereas others, inserted for a short while, can be used to take samples of eggs.

population cage

We would expect AR and CH to stabilize at 50 per cent but we have already seen (Table 17) that this does not happen. Note, however, that when a temperature of 15°C is used, ST and CH do stabilize out at 50 : 50.

QUESTION Apart from temperature, can you suggest other components of the environment that may be responsible for maintaining the chromosome III polymorphism in *D. pseudoobscura*?

ANSWER There are numerous possibilities of which humidity, atmospheric pressure, availability of food and pest pressure are the most likely.

There is some evidence about the effect of humidity. At 100 per cent relative humidity, CH is more viable than either AR or ST. At 92 per cent relative humidity ST is the most viable, whereas at zero relative humidity, AR is the most viable.

At Piñon Flats the annual cycle in the relative frequencies of the ST and CH chromosome can be represented as in Figure 19. Clearly, something is happening in the spring to change the selective advantage of the ST arrangement into a relative

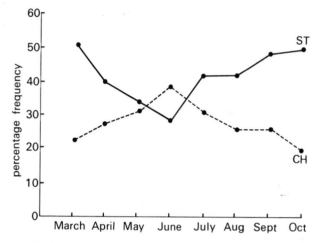

Figure 19 Changes in frequency of Standard and Chiricahua inversion types of chromosome III of *D. pseudoobscura* from March to October at Piñon Flats, San Jacinto Mountains, California.

disadvantage. Birch wondered whether *food* was important. He suggested that yeasts accelerate their reproduction in the spring, and therefore there will be little larval competition for food at that stage. Later in the spring, however, there will be an increase in the number of larvae and this will lead to competition between them, both for food and for space.

In experiments designed to test these ideas he showed that when an abundance of food was supplied to an ST–CH population-cage experiment at 25°C, CH was at an advantage!

QUESTION How does this information help to explain the cycle at Piñon Flats?

ANSWER In early spring when there are more yeasts, there is little larval competition and consequently CH increases. In late spring and summer larval competition builds up, placing CH at a disadvantage relative to ST, and so it declines in frequency.

What happens biochemically is another matter and is beyond the scope of this Course.

An enormous amount of work on the natural populations of *D. pseudoobscura* has been done by Dobzhansky, his associates and his pupils. There is now nearly 40 years' worth of data collected from populations that have been visited at regular intervals. In that time there have been changes, but there have also been many examples of marked stability. One of the most noticeable changes is in the frequency of the chromosome arrangement known as Pikes Peak (PP). In California it rose rapidly in frequency from zero per cent during the period 1936–40 to a peak of 12.9 per cent in one population in the Panamint Mountains in 1963, falling away again in most populations subsequently. Concomitant with this decline in the past ten years there has been an increase in the frequency of TL. The time-scale of the increase and decrease in the frequency of PP parallels the increase and decline in the use of insecticides, particularly DDT, but a causal relationship, although mooted, has yet to be established.

In conclusion, this work has shown that the selective value of a particular inversion chromosome varies according to the geographical region and also the seasonal variations within a single locality. Furthermore, the inversions protect co-adapted gene complexes from recombination at meiosis (crossing-over is suppressed), so that an integrated group of genes can be built up and thus form a supergene: the supergene is subject to selection rather than the individual genes. In *D. pseudoobscura*, it appears that the different supergene complexes maintained by the different inversion chromosome arrangements have become individually adapted to different facets of the environment and are thus maintained in a balanced polymorphic state.

You should now attempt SAQ 8.

10.1.5 Systems promoting outbreeding

> **ITQ 10** (a) Define inbreeding. (b) Describe the primary genetic consequences of inbreeding.

Genetic variation is the raw material of evolution by natural selection. The production of genetic variation has two main advantages. It allows the exploitation of a greater variety of ecological sub-niches and the colonization of marginal habitats. In this way the environment is more fully exploited. If the environment changes drastically, there must be genetic variation to meet the demands of these changes. On the other hand the release of too much variation may lead to the production of a variety of disadvantageous genotypes, which will be weeded out by natural selection. A population must, therefore, strike a compromise between the conservation and the release of genetic variation.

Remember that although the primary origin of genetic variation lies in gene mutation, the *major* source of variation is through recombination, producing new genotypes at meiosis and at fertilization. The amount of variation is, therefore, under the control of the *sexual* system.

It should also be remembered that a sexual system doesn't merely imply males and females. This sexual dimorphism is only an outbreeding mechanism superimposed on a sexual system and we shall, therefore, refer to a situation with two mating types, for example males and females, as *dioecy*. A sexual system implies the regular alternation of diploid and haploid phases. The haploid phase is preceded by meiosis, and succeeded by fertilization to give the diploid phase.

dioecy

The amount of recombination and, therefore, the amount of potential variation held in the gene pool is under the control of the meiotic system. The release of this variation is controlled by the breeding system. We can distinguish two extremes:

(a) A high degree of inbreeding, which leads to homozygosity, reduction in variability, and a reduced adaptability to future environmental changes.

(b) Outbreeding, which leads to heterozygosity and greater variation. This increases the potential for adaptation in the future, should the environment change.

A balance is usually maintained between these two extremes. In nature, however, far more systems exist to promote outbreeding than to reduce it, and some of these systems are of particular interest to us because they involve polymorphisms.

Dioecy

Dioecy is the most important mechanism for promoting outbreeding and its characteristic is that the two gametes participating in any fertilization event must be from individuals of different mating type, or sex. Dioecy is the rule in complex animals, but it is rather uncommon in plants and simple animals.

A regular feature of dioecious systems is that each cross is a backcross as far as mating type is concerned. In humans and *D. melanogaster*, for example, where males have an XY sex-chromosome constitution, and females an XX constitution, each mating must be between XX and XY individuals, and a 1 : 1 ratio of XX and XY is expected among the progeny. In grasshoppers, however, the females have two X-chromosomes, but males have only one X and no Y-chromosome. Half of the male gametes, therefore, will carry an X and the other half will carry no sex chromosome. Every cross is again a backcross, producing a 1 : 1 ratio of XX and XO among the progeny.

Figure 20 The white campion, *Silene alba*. (a) Male (staminate) plant and (b) female (pistillate) plant. The flowers are cut open to show the absence of any female development in (c) and the absence of any male development in (d).

About 2 per cent of the British flora are dioecious; that is, they bear male and female flowers on separate plants. There is good genetic evidence that these systems are controlled by sex chromosomes and that dioecy is accomplished in the same ways in the plant kingdom as in the animal kingdom. In the red and white campions, *Silene dioica* and *Silene alba* (see Fig. 20), for example, the X and Y-chromosomes can be cytologically distinguished. Males have an XY and females an XX sex-chromosome constitution, as in humans. It is the presence or absence of the Y-chromosome that determines whether a plant is male or female.

The sex chromosomes of *Silene* are shown in Figure 21. The two chromosomes differ in size, the Y being considerably larger, and for most of their length they are not homologous. In the male the only parts of the chromosomes that pair at meiosis are the homologous segments labelled IV in the diagram. Segments I, II and III comprise the differential part of the Y-chromosome and segment V is the differential part of the X-chromosome. These segments are therefore termed *differential segments*.

The genes within each of the differential arms are completely linked because these segments do not pair at meiosis in the male. In the female, of course, there is complete homology between the two X-chromosomes and so recombination in the differential segments will occur normally. The mechanism of sex determination in *Silene* was discovered as a result of studies on plants in which part of the Y-chromosome was missing.

Plants with segment I missing are hermaphrodite—that is, the flowers show both male and female fertility—and they can, indeed, be self-fertilized. It is clear, therefore, that in the normal male plant this segment I of the Y-chromosome contains genes that suppress female sexual development.

Plants lacking segment I have normal vegetative and male sexual development. Consequently, we may infer that there are few other genes in this region of the Y-chromosome.

Plants with a Y-chromosome lacking the differential segment III and the homologous segment IV are sterile males. The early stages of anther development and meiosis in the pollen mother cells are normal, but the anthers then degenerate. This implies that segment III contains genes that control the later stages of male sexual development.

differential segment

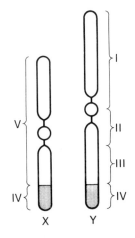

Figure 21 The sex chromosomes of *Silene*. The shaded segments pair at meiosis (they are known as *pairing segments*); the unshaded differential arms do not pair.

> **ITQ 11** We know that XX plants (that is, those containing no Y-chromosome at all) are female. Suggest, therefore, the location of the genes that control the early stages of anther development in males.

Finally, it has been shown, as expected, that the genes determining normal female development are carried in segment V—the differential part of the X-chromosome.

Dioecy in *Silene* can then be seen to be determined by a simple chromosomal switch system involving complete linkage between genes that suppress female development and genes that control the early and late stages of male development. Westergaard* has compared this system

> to a machine which has two potential functions, male production and female production. A machine can be stopped in two different ways, either by applying the brakes or by removing some essential wheels without which the machine cannot work. The Y-chromosome serves both purposes. When it is present the brakes are pulled on the female-producing part of the engine, and the essential wheels are in place in the male-producing part. The result is a male plant. When the Y-chromosome is absent the brake is released from the female-producing part, but at the same time some essential wheels are removed from the male-producing part, the result is a female. The efficiency of this principle is due to the complete linkage between the 'female brake' and some essential 'male wheels'.

The maintenance of the sex ratio

In humans the current *sex ratio* at reproductive age in England is not 1 : 1; it is nearer 109 males to 100 females. But this has not always been the case. During most of the last century in England there was approximate equality in the sex ratio.

sex ratio

> QUESTION What could have brought about this change?

> ANSWER The frequency of still-births is higher among male babies than among females and it is well known that male children have a higher infant mortality rate. In other words, males are biologically the weaker sex. Modern advances in medicine have improved the prospects for male survival and consequently more male children are surviving to reproductive age.

> QUESTION From the data available in the three sentences before the previous question, estimate the likely sex ratio of live births among the English population one hundred years ago.

* Extract from Westergaard, M. (1958). The mechanism of sex determination in dioecious flowering plants. *Adv. Genet.* **9**, 217–94.

ANSWER 109 ♂ : 100 ♀

Examination of the records shows that the sex ratio at birth was indeed approximately 109 : 100. This suggests that the sex ratio of live births in humans has been adjusted in such a way that an optimum sex ratio is achieved post puberty. Perhaps the optimal post-puberty sex ratio is still 1 : 1 today, but sufficient time has not yet elapsed for there to have been the necessary adjustment in the live-birth sex ratio required to compensate for the improved life expectancy of males.

But how can the sex ratio be adjusted through evolution? This has proved to be a difficult problem because why should an individual who produces progeny with an unequal sex ratio differ in fitness from an individual who produces equal numbers of males and females? The chromosomal sex-determining mechanisms certainly ensure an approximation to an equal sex ratio. In some organisms, however, some autosomal genes have been found to influence the sex ratio, so genetic variation is present on which natural selection could operate.

Darwin was aware of the problem, but was baffled by it. He wrote in *The Descent of Man*: 'I now see that the whole problem is so intricate that it is safer to leave its solution to the future'. The answer was provided, some sixty years later, by Fisher.

The important feature of Fisher's argument is that he considered, not only the progeny, but also the grand progeny. If a population contains an excess of males (for argument's sake), each female will have, on average, more progeny than each male. Any parent, of either sex, whose genotype is such that the sex ratio among their progeny favours females will, therefore, contribute more to the subsequent, or grand progeny, generation. Any genes leading to an adjustment of the sex ratio towards equality will, therefore, increase under natural selection until a 1 : 1 sex ratio is achieved. The same argument applies in reverse if there are more females than males. Any deviations from an equal sex ratio will tend to be corrected.

Returning now to *S. alba*, some interesting suggestions have been made as to how the sex ratio is maintained in this species. It has been found that when the stigmata of a female plant receive a small amount of pollen then the progeny show a sex ratio of approximately 1 : 1. If, however, the pollen is in excess, then the progeny show a greater proportion of female plants. This can be explained by supposing that pollen tubes carrying an X-chromosome grow faster down the style than those carrying a Y-chromosome. Thus if a large quantity of pollen is applied to the stigmata most of the ovules will be fertilized by the faster-growing pollen tubes, which carry an X-chromosome. An excess of female progeny will result. Consequently pollen will become sparse. In these circumstances, sufficient ovules will remain unfertilized by X-carrying pollen tubes for more of the slower-growing Y-carrying pollen tubes to effect fertilization. The result will be a sex ratio in the progeny that is nearer to equality. Thus natural selection can adjust the sex ratio, because if a population contains too many males there will be a high frequency of crosses involving excess pollination. This leads to an excess of females in the progeny. On the other hand, if the population contains few males, most crosses will involve only sparse pollen and females will not be favoured at the expense of males among the progeny. This system is balanced slightly in the direction of females. Observations on natural and experimental populations suggest that the optimal sex ratio, as reflected in seed output, does indeed differ from 1 : 1 in favour of females.

We mentioned earlier that only about 2 per cent of species in the British flora are dioecious and bear male and female flowers on separate plants. What about the other 98 per cent of species? About 5.4 per cent have separate male and female flowers, but they are borne on the same plant. These are known as *monoecious* species. A very small proportion (0.6 per cent) have other combinations of sex organs; for example, some, known as *gynodioecious*, have plants of two kinds: some plants bear only female flowers, and the remainder have *hermaphrodite* flowers; that is, each flower has full male and female fertility. The remainder of the British flora, in fact the majority of species, are wholly hermaphrodite (see Fig. 22).

monoecy

gynodioecy
hermaphrodite

Hermaphrodites do not necessarily practise regular self-fertilization because other systems have evolved to reduce inbreeding. Thus many species are *protandrous* (the anthers release pollen before the stigmas in that particular flower are receptive) or *protogynous* (the stigmas are receptive before the anthers shed pollen in that particular flower). But self-fertilization can still occur when pollen is transferred to the stigmas of another flower on the same plant. Hermaphrodites do, however, have other tricks up their styles—incompatibility systems.

protandrous

protogynous

450

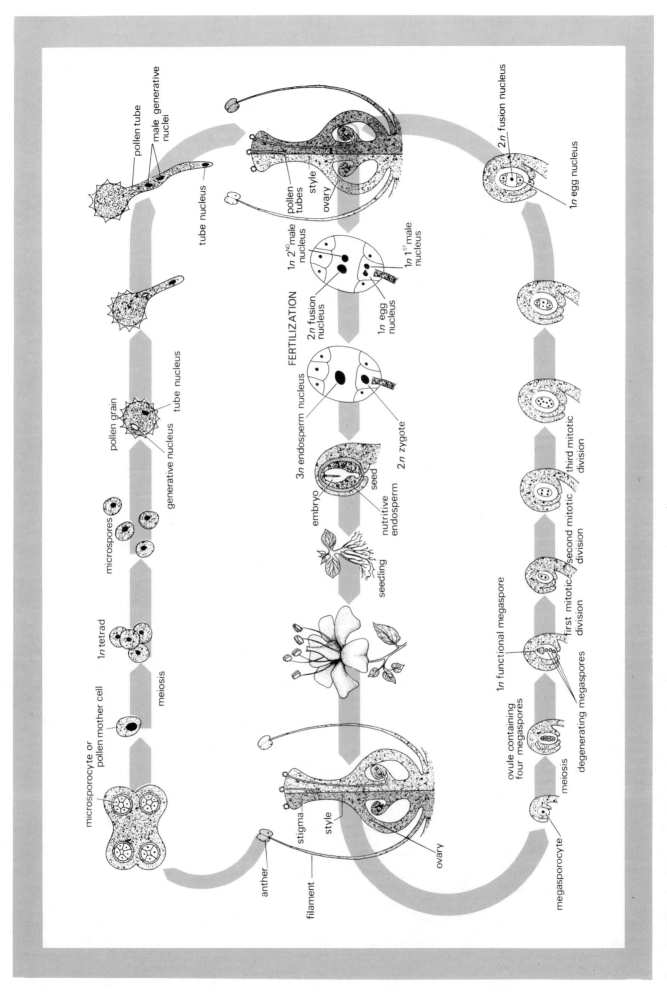

Figure 22 Life cycle of a hermaphrodite flowering plant. The zygote develops into a new diploid plant that bears the flowers in which these events are repeated.

10.1.6 Incompatibility systems in higher plants

Many species of plant with hermaphrodite flowers fail to set seed when they are self-pollinated. Furthermore, even cross-pollinations between some plants of the same species may result in no seed production, whereas cross-pollinations between other pairs of plants may be successful. These species are said to possess *incompatibility systems*.

Studies on the sweet cherry, *Prunus avium*, in the 1920s established that self-pollination within a tree failed to produce fruit. Moreover, cross-pollination between some trees failed as completely as self-pollination, but crosses involving other pairs were completely fruitful. The results with reciprocal crosses were always the same. It was soon established that the various varieties of commercial and ornamental sweet cherries could be arranged into about 16 groups. No two varieties within a group could be crossed successfully, whereas pairs of varieties belonging to different groups were cross-fertile. A successful orchard must, therefore, contain at least two different varieties belonging to separate compatibility groups.

It was clear that the incompatibility between varieties belonging to different groups must reflect some pollen–style interaction. If was found that when incompatible varieties were crossed the pollen tubes grew down the style so slowly that fertilization was never effected. Compatible pollinations were of two types. Either all the pollen tubes grew well or, sometimes, sufficient showed normal growth to fertilize all available ova but a quantity of pollen tubes showed retarded growth and did not effect fertilization.

Parent–offspring crosses were always found to be of this latter type when the 'female' plant was the backcross parent. Herein lay the clue to the genetic determination of the pollen–style interaction. It was suggested that the plants that produced the two types of pollen, fast- and slow-growing, were heterozygous at a particular locus (say $S_1 S_2$). Because pollen grains are the products of meiosis and, therefore, haploid, they will be of two types, S_1 and S_2. The fact that the parent–offspring crosses are half-compatible and self-pollinations are fully incompatible suggests that pollen tubes do not grow effectively down the style of a plant that has in its diploid genotype an *S allele* the same as that carried by the pollen. Predictions based on this model were quickly confirmed by crosses in a variety of species.

We may consider three different types of mating—those where the parent plants have both, one or no *S* alleles in common (see Fig. 23).

(a) Both *S* alleles in common. For example

$$\begin{array}{ccc} \female & & \male \\ S_1 S_2 & \times & S_1 S_2 \end{array}$$

$$\text{style } S_1 S_2 \qquad \tfrac{1}{2}S_1 : \tfrac{1}{2}S_2 \text{ pollen}$$

Both types of pollen grain carry an *S* allele that is also represented in the style genotype. The cross is, therefore, *fully incompatible*. Self-pollinations are of this type.

(b) One *S* allele in common. For example

$$\begin{array}{ccc} \female & & \male \\ S_1 S_2 & \times & S_2 S_3 \end{array}$$

$$\text{style } S_1 S_2 \qquad \tfrac{1}{2}S_2 : \tfrac{1}{2}S_3 \text{ pollen}$$

Only the S_3 pollen grains can effect fertilization and the cross is therefore *semicompatible*. Parent–offspring matings are of this type.

(c) No *S* alleles in common. For example

$$\begin{array}{ccc} \female & & \male \\ S_1 S_2 & \times & S_3 S_4 \end{array}$$

$$\text{style } S_1 S_2 \qquad \tfrac{1}{2}S_3 : \tfrac{1}{2}S_4 \text{ pollen}$$

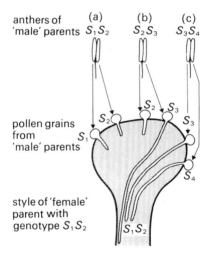

anthers of 'male' parents

pollen grains from 'male' parents

style of 'female' parent with genotype $S_1 S_2$

Figure 23 Pollinations where the parent plants have (a) both, (b) one and (c) neither *S* allele in common.

Both types of pollen grain can effect fertilization because neither carries an S allele that is represented in the style genotype. This cross is known as *fully compatible*.

> **ITQ 12** (i) What are the genotypes and their proportions among the progeny of the three types of cross described above?
>
> (ii) What are the compatibility relationships between these progeny and each parent?
>
> (iii) What are the compatibility relationships when the progeny are sib-mated?
>
> (iv) What is the outcome of the cross $S_1S_1 \times S_2S_2$?
>
> (v) What is the minimum number of S alleles that a population must contain?

Any new S allele arising in a population by mutation has an immediate selective advantage because the individual carrying it will be fully or semi-compatible with all other plants in the population. As the frequency of the allele increases, its selective advantage will wane because incompatible pollinations involving this allele will also increase in frequency. If an S allele is present at high frequency, it will suffer a selective disadvantage because pollen grains carrying it will often land on styles with which they are incompatible. It is not surprising, therefore, that natural populations of species that possess these incompatibility systems are found to be highly polymorphic at the S locus. Red clover populations, for example, have been found to contain in excess of 20 different S alleles.

There is a particularly remarkable population of the evening primrose, *Oenothera organensis* in the Organ Mountains (New Mexico). Although this population contains only about 500 plants, it is estimated that its incompatibility system is maintained by no less than 45 different S alleles. This is a remarkably large number of S alleles for such a small population, and mathematical arguments show that it is necessary to postulate a mutation rate to new or extinct S alleles of about 10^{-3} per generation in order for this number of alleles to be maintained in equilibrium. This would be an unusually high mutation rate. However, subsequent experiments by Lewis on spontaneous and radiation-induced mutation have shown that the only mutations of the S gene are loss mutations to self-fertility alleles. This finding is of considerable interest because it points to an unusually *low* mutation rate to new S alleles, probably less than 10^{-8} per generation. The generally favoured explanation for the high number of S alleles in *O. organensis* is that the population was at one time very much larger, but since the beginning of the nineteenth century there has been extensive grazing in the area and this has led to a drastic reduction in population size. If this explanation is correct we can expect a gradual decline in the number of S alleles over the next few decades.

Because the pollen reaction is determined by its own haploid genotype, the incompatibility systems that we have so far discussed are called *gametophytic*. There are, however, other examples where the pollen parent's diploid phenotype determines the reaction of the pollen. These are therefore called *sporophytic* systems and are found, for example, in cabbages and their allies (*Brassica* species). A pollen grain will not effect fertilization if the diploid genotype of the style on which it lands is the same as the diploid genotype of the plant that produced the pollen. It is probable that the pollen reaction is determined via the paternally originating cytoplasm of the pollen grain. Incompatibility is again determined by a single S gene with many alleles. But both pollen and style reactions depend on diploid genotypes, and alleles may act independently, or show dominance, in both the pollen and the style. Furthermore, the dominance relationships between alleles in the pollen need not necessarily be the same as those between the same alleles in styles. Sporophytically determined systems can therefore be considerably more difficult to unravel, and the first one was not fully understood until 1950.

gametophytic incompatibility system

sporophytic incompatibility system

In *Crepis foetida* (Fig. 24), for example, S_1, S_2 and S_3 are all dominant to S_4, but act independently of each other in the style. In the pollen, however, S_1 and S_2 are independent of each other, but both are dominant to S_3, which is, in turn, dominant to S_4. This situation is indicated diagrammatically in the margin.

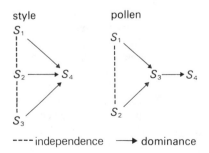

> **ITQ 13** What are the genotypes, and their proportions, of the progeny from reciprocal crosses between *C. foetida* with genotypes S_1S_3 and S_3S_4?

Figure 24 Stinking Hawk's-beard, *Crepis foetida*.

In the answer to the last ITQ you can observe the two important diagnostic features that distinguish sporophytic from gametophytic systems:

(a) The outcomes of reciprocal crosses need not be the same.

(b) Homozygotes may arise and are, in fact, a regular feature of sporophytic systems.

Notice also that because pollen reaction is determined sporophytically all pollen grains have the same reaction. Crosses are therefore either fully compatible or fully incompatible; semi-compatible crosses are not possible.

These sporophytic systems are determined by a physiological reaction, between the style and the germinating pollen tube. It is not possible to tell the genotype merely by inspecting the phenotype of a plant. They are therefore called *homomorphic* in-compatibility systems. Other sporophytic systems differ in that superimposed on the physiological reaction are morphological differences that reinforce the incompati-bility mechanism. These systems are called *heteromorphic*, and here it *is* possible to determine the genotype by inspecting the phenotype of a plant. We shall finish our study of incompatibility mechanisms with a quick look at these systems, which are particularly interesting because they result in a situation that is not unlike dioecy even though they are fully functional hermaphrodites.

homomorphic incompatibility system

heteromorphic incompatibility system

Heterostyly in the primrose

It has long been known that members of the primrose family, for example the common primrose and the cowslip, show a rather unusual feature. Natural popula-tions contain two types of plant, often in approximately equal proportions. On some, the style is long, so that the stigma is positioned near the opening of the corolla tube, but the anthers are situated at a lower level about half way down the tube. These plants are called *pins* (Fig. 25a). On other plants, however, the relative positions of stigma and anthers are reversed so that a short style bears a stigma about half way up the corolla tube and the anthers form a fringe around the opening of the tube. These plants are called *thrums* (Fig. 25b). This difference in the length of the style is termed *heterostyly*.

pin

thrum
heterostyly

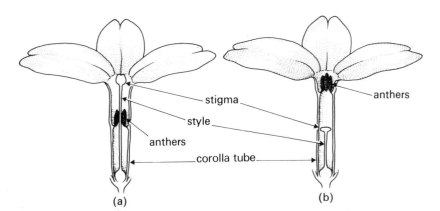

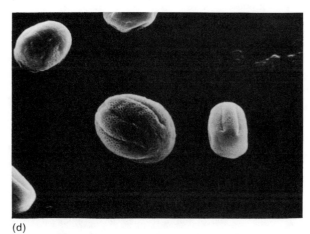

Figure 25 Common Primrose, *Primula vulgaris*. (a) Half of a pin flower, (b) half of a thrum flower, (c) pin and thrum stigmatic surfaces (the pin is the larger) and (d) pin and thrum pollen (the pin is the smaller); magnifications: (c) ×29, (d) ×740.

The implications of this dimorphism in floral structure were, however, a complete mystery until the results of some of Darwin's studies were published in 1877, in his book *The Forms of Flowers*. He found, first of all, that primroses and cowslips did not set seed if insects were prevented from visiting them. This implied that they were self-sterile, and this was confirmed by the failure of artificial self-pollinations to produce seed. He then artificially transferred pollen from one flower to another on a different plant, again preventing access to insects, with the results summarized in Table 19.

Table 19   The results of Darwin's artificial pollinations in the primrose

Pollination		No. of flowers pollinated	No. resulting in seed	Average number of seeds/pollination
Seed parent	Pollen parent			
thrum × pin		8	7	56.9
pin × thrum		12	11	61.3
thrum × thrum		18	7	7.3
pin × pin		21	14	34.8

It is quite apparent that crosses between plants that bear the same type of flower are relatively infertile compared with crosses between pins and thrums. Darwin regarded heterostyly as strong evidence in support of his theory of evolution. Self-fertilization was thought to be usually disadvantageous in the long run and heterostyly was presumed to have evolved through natural selection as a method of ensuring cross-fertilization. The insects that visit the flowers in natural populations have a long proboscis, which they extend down the corolla tube in search of nectar. Pollen from pin and thrum plants, therefore, will collect in different places on the proboscis — pin pollen towards the distal, or free, end and thrum pollen towards the proximal, or head, end. The pollen is then in the appropriate position to maximize the chances of cross-fertilization, pin pollen to thrum stigma and vice versa.

This is not, however, the end of the story. It was found that the legitimate pollinations thrum × pin and pin × thrum yield a 1 : 1 ratio of thrums to pins. But the illegitimate artificial pollinations produced 3 : 1 ratios of thrums to pins from thrum × thrum, and pins only from pin × pin pollinations.

ITQ 14 Suggest the genetic basis of the difference between the thrum and pin characters.

Notice that legitimate pollinations are always between heterozygote and recessive homozygote so that we have a perpetual backcross system. The situation is not, then, unlike dioecy, but with heterostyly any individual can produce both male and female gametes.

The difference between pins and thrums, however, is more complex than a cursory glance reveals. The thrum stigma is flattened and bears short papillae, whereas the pin stigma is globose and has papillae about five times as long (Fig. 25c). At one time it was thought that this difference in pollen size (Fig. 25d) was the critical factor in determining incompatibility. Thrum pollen, being larger, can produce longer pollen tubes, which therefore reach down the long pin style more easily than the short pin pollen tubes.

> QUESTION There is a snag in this argument. Can you spot it?

> ANSWER The theory implies that thrum pollen, with its long pollen tubes, would grow effectively down the short thrum style. We have seen in Table 19 that whereas pin × pin crosses are partly successful at producing seed, thrum × thrum crosses are almost completely unsuccessful.

We now know that the important attribute of pin and thrum flowers is not morphological, but rather a physiological incompatibility reaction between pollen and style. The morphological differences merely reinforce the physiological incompatibility mechanism. Thrum pollen does germinate on a thrum stigma, but does not penetrate the surface. Pin pollen both germinates on, and penetrates, a pin stigma. But, compared with thrum pollen, pin pollen grows relatively slowly down the pin style and therefore reaches the egg nuclei too late to effect fertilization. Thus the critical incompatibility reaction is similar to the one we find in the more typical S-allele systems, and, because the thrum pollen reaction does not show segregation, the system is of the sporophytic type.

Careful genetic analysis has shown that the system is not, in fact, simple. The S gene is really a supergene of at least seven loci, the dominant alleles of which determine the thrum area of stigmatic surface, style length, stigmatic papilla length, style incompatibility, pollen incompatibility, pollen size and anther position. The loci are very tightly linked and, because crossing-over is rare, behave effectively as a supergene in their transmission, but not, we suspect, in genetic transcription.

> QUESTION Given that the physiological incompatibility mechanism is efficient, what can be the advantages of superimposing on it the morphological floral differences?

> ANSWER Pollen is carried on insects' probosci at a position where it is most likely to be transferred to a stigma of the appropriate compatible type. This will ensure that the majority of the pollen landing on a stigma will be compatible, and gamete wastage is thereby effectively reduced.

> **ITQ 15** If cross-fertilization fails, pins sometimes self-fertilize but thrums remain sterile. What can be the advantage of greater self-sterility in thrums than in pins?

In the last two Sections we have cited a number of polymorphisms related to systems that promote outbreeding, in particular dioecy and the incompatibility systems in hermaphroditic flowering plants. Outbreeding reduces the frequency of matings between close relatives and affects the relative frequencies of heterozygotes and homozygotes at all loci in the genotype. Disassortative mating is the tendency for mating to occur between individuals of unlike phenotypes for a particular character. We pointed out in Section 9.4.4 that disassortative mating will only affect the distributions of homozygosity and heterozygosity for those loci involved in the determination of the particular character by which mating choice is effected and other genetically linked loci. We shall now consider a polymorphism that is related to disassortative mating, and, as we shall see, the distinction between outbreeding and disassortative mating is fundamental. Polymorphisms maintain the outbreeding systems, but disassortative mating maintains the polymorphism. The cause-and-effect relationship is reversed.

10.1.7 Disassortative mating in *Panaxia dominula*

The Scarlet Tiger, *Panaxia dominula*, is a daytime flying moth, which is cryptic at rest, but when in flight displays vivid black and red colouration. These moths are distasteful to birds and are therefore *aposematic*; that is, they possess the dual quali-ties of noxious properties and warning colouration. Colonies of *P. dominula* occur in scattered localities throughout southern England and one of these, in a private estate at Cothill, Berkshire, can perhaps claim to be the most intensively studied natural population of any animal. This colony has attracted considerable interest because it is polymorphic for colour patterning (Fig. 26). The difference is deter-mined by a pair of alleles, B^D and B^M. The common variety, *dominula*, is homozygous $B^D B^D$, and displays the typical colour pattern. The rare homozygote $B^M B^M$ is called *bimacula* and shows a great increase in the area of black pigmentation on the wings. The heterozygote $B^D B^M$, known as *medionigra*, shows intermediate expression and can be distinguished from the two homozygotes. It is possible, therefore, to estimate the allele frequencies by direct gene counting from a sample, as in the answer to ITQ 2. This polymorphism differs from those that we have already considered where populations were regularly polymorphic. It is only in the Cothill population that the B^M allele achieves a frequency greater than that of an occasional mutant. It is, there-fore, of great interest to determine the nature of the selective forces maintaining the polymorphism at Cothill.

Ford and his colleagues have studied the frequency of the B^M allele and also the population size of the Cothill colony from 1939 to the present. The frequency of B^M fluctuates, but it is always rare, usually around 0.02–0.04. The population size also fluctuates, but usually comprises between one and ten thousand individuals.

In 1952 Sheppard recorded the phenotypes of copulating moths in population-cage experiments. Each trial involved three moths: one of a particular sex and genotype and two of the alternative sex, one with the same and one with a different genotype. A trial might, for example, comprise male *dominula*, female *dominula* and female *medionigra*. Although the results from different types of trial are somewhat hetero-geneous (see *STATS*, Section ST.4.6), the over-all outcome was that out of a total of 150 trials 97 resulted in matings between unlike genotypes and only 53 in matings between like genotypes. This was caused by females preferring to mate with males unlike themselves. The males, however, were indifferent! In nature, females are in-seminated only once but a male may fertilize many females. Males tend to assemble in some numbers around a virgin female and if there is a tendency towards dis-assortative mating, which seems not unlikely, then rare phenotypes in males will be at an advantage. As the frequency of the B^M allele rises, this advantage will wane.

Other experiments support this hypothesis. For example, 4 000 eggs from backcross matings between *dominula* and *medionigra* were scattered in 1951 in a suitable habitat at Hinksey (Oxford), where *P. dominula* was absent. The frequency of the B^M allele, initially 0.25, stabilized at about 0.06 to 0.08 over the next ten years (see Table 20).

In another locality, Sheepstead Hurst (Berkshire), there was a population that lacked the B^M allele. In 1954 eggs from 50 *medionigra* back-crossed to *dominula* were scattered in the locality, introducing the B^M allele at an estimated initial frequency of 0.000 2. By 1961 the frequency had risen to about 0.02 and has since fluctuated around this value. There is, therefore, some evidence that *medionigra* is at some advantage while it is rare, but suffers a net selective disadvantage when common. Incidentally, over the years at Cothill, *medionigra* has evolved to look more like *bimacula*. At Sheepstead Hurst, however, *medionigra* has become more like *dominula*. We have some evidence, therefore, that dominance has evolved in different directions in these two colonies (see Section 9.7.2 and SAQ 5).

In conclusion, the evidence does suggest that the polymorphism is being maintained, at least in part, by disassortative mating. There must be other selective forces operat-ing against *medionigra*, because its frequency appears to stabilize at a low value. There is some evidence that *medionigra* larvae suffer a higher mortality than do *dominula*. It is also possible that the change in colour pattern resulting from the possession of the B^M allele upsets the association between warning colouration and noxious properties which is necessary to deter predators.

You should now attempt SAQ 9.

aposematic

Figure 26 Scarlet Tiger moth, *Panaxia dominula*. (a) The typical (*dominula* form) and (d) the homozygote *bimacula*. (b) and (c) are examples of the heterozygote *medionigra*. There is considerable variation in the expression of the *medionigra* allele in the heterozygotes, but the name refers to the reduction or absence of the white spot in the middle of the forewing.

Table 20 The frequency of the B^M allele in an artificial population

Year	Frequency of B^M allele
1951	0.25
1952	0.15
1959	0.065
1960	0.062
1961	0.073

10.1.8 Summary of genetic polymorphism

We could continue giving different examples of genetic polymorphism more or less *ad nauseam*, but we think we have described enough for you to see the variety that has been studied and to gain a 'feel' for the way in which the problems are tackled. You should now have no problem with the next ITQ!

ITQ 16 Which came first, the chicken or the egg?

In Section 10.1, we posed two questions. How widespread is genetic polymorphism? Is it important in evolution? From the various examples we have discussed it should now be clear that genetic polymorphism is a common phenomenon in most if not all species and can be maintained in a variety of ways. Let us summarize, therefore, by listing some of the ways by which genetic polymorphisms can be maintained, considering both situations in which there are *constant* selective values and those in which there are *variable* selective values. The list is based on one compiled by Williamson in 1958, but it concentrates on those examples already discussed and on others that will be considered later in these Units. But beware! Because the basic model suggests that we should look for a balance of selective forces when we study a genetic polymorphism, it should not be surprising to find that *more than one of the mechanisms* listed may be involved in the maintenance of a particular polymorphism. Furthermore, no known habitat is entirely uniform and so no individual can be subjected to exactly the same selection as any other.

1   With constant selective values

(i) Simple segregation.
(a) Heterozygous advantage—sickle-cell haemoglobin;
(b) Systems of multiple alleles—shell colour in the snail *C. nemoralis*, blood groups in humans.

(ii) Interacting with other polymorphic systems.
(a) Sex linked;
(b) Associated with sex dimorphism but not necessarily sex linked—absence of tails in mimetic female swallowtail butterflies of the genus *Papilio*.

2   With variable selective values

(i) Selection varies with space or time.
(a) Clines—industrial melanism in *B. betularia*, papillate seeds in *S. arvensis*;
(b) Cyclical seasonal changes—chromosomal polymorphism in *D. pseudoobscura*.

(ii) Frequency-dependent selection depending on interactions between individuals of the same species.
(a) Sex and similar outbreeding systems;
(b) Disassortative mating—mating preference in the scarlet tiger moth, *P. dominula*.

(iii) Frequency-dependent selection depending on other, particularly biotic, variables.
(a) Commoner form at a disadvantage—*C. nemoralis*.

So, genetic polymorphism is widespread, it can be maintained in a variety of ways and it can tell us a lot about evolutionary processes, but it is doubtful whether it can really be at the basis of evolution. The problem is that the examples we have considered have involved more or less *stable* situations, that is, balanced polymorphisms, and have been so for many years. Admittedly there was a major change in the frequency of the melanic form of *B. betularia* during the last century, yet as soon as the selection pressure was relaxed, there was a tendency to revert back to the original state. An evolutionary process was involved, but it did not go to completion. Thus it would seem that genetic polymorphism is a consequence of fluctuating selection in a population, different morphs having different relative fitnesses from time to time and from place to place, and yet when we look at the whole there is apparent stability.

At the end of the programme notes associated with TV programme 9 we asked the question 'Is the genetic polymorphism itself of importance to the species?'. With the snails *C. nemoralis* and *C. hortensis* there is no evidence at all that the species would become extinct if it were to become monomorphic overnight. Indeed there is no evidence to suggest that these snails would not occur in the same places and in the same numbers as at present.

We have, however, discussed another situation (p. 445) which at first sight does suggest that the polymorphism is advantageous to the species. We saw that experimental populations of *D. pseudoobscura* produced a larger number of individuals and a greater biomass when polymorphic for AR and CH than when monomorphic for either of the chromosome arrangements. But this does not prove that a polymorphic population is better adapted to *wild* habitats than a monomorphic population. All it shows is that the heterozygotes are at an advantage over the two homozygotes *when the population is polymorphic*. We shall remind you of what we stated on p. 410: we have no means of measuring absolute fitness. We cannot, therefore, compare monomorphic with polymorphic populations. Let us return to sickle-cell haemoglobin again. You could argue that because the heterozygote Hb^AHb^S is the most fit genotype, human populations survive in those regions of the world where falciparum malaria is common only because they are polymorphic. On the other hand, the low viability of Hb^SHb^S individuals has *nothing* to do with malaria. Indeed, if a way could be found to keep Hb^SHb^S individuals alive so that they had a relative fitness of 0.9 instead of 0.2, the Hb^S allele would tend to replace the Hb^A rapidly. Eventually the human population would reach a new equilibrium level with the frequency of Hb^AHb^A individuals being only 8 per cent.

> QUESTION Can you think of two types of genetic polymorphism that we have described earlier that could not become monomorphic without leading, almost inevitably, to the extinction of the species?
>
> ANSWER (a) Sexual dimorphism; (b) *S*-allele incompatibility systems.

We can conclude, therefore, that a population will be polymorphic for characters other than those concerned with the breeding system because the different morphs have different selective values and not because the polymorphism itself necessarily has any intrinsic value.

We know almost nothing about the origin or the history of any of these genetic polymorphisms. It does not require any great inspiration to guess that mutation is the basic origin of variation, but no one has observed this directly in the wild.* All the evidence we have is circumstantial and leads us to infer that mutation recurred some time previously. Furthermore we have no idea how much of the variation we see on the ground reflects selection now and in the past or how much has resulted from other processes that cannot be quantified. How much, for example, is the variation in the distribution of human ABO blood groups the result of selection and how much is it a consequence of invasion, forced migrations and lightning raids ending in the rape of the local women? As Watkin has argued, these possibilities must be taken into account if we are to explain why the I^O allele in humans varies from 55.9 per cent around Pembroke Docks to over 75 per cent in the region around Harlech and Penrhyndeudraeth, whereas in the rest of England and Wales the frequencies lie in between these extremes. We shall return to such problems when we consider the rate of evolution (Section 10.6) and genetic drift (Section 10.5).

Consequently we must regard the theoretical models we outlined in earlier Sections as being too simple. Don't worry, because apart from a consideration of frequency-dependent selection we are not going to *extend* the models in this Course, although we shall be *using* some of them for calculations. We can say now, however, that it is not yet possible to describe any genetic polymorphism in part, let alone as a whole, in terms of an adequate model—not even sickle-cell anaemia.

10.2   Frequency-dependent selection   [B]

In Sections 10.1.1 to 10.1.7 we have met a large number of polymorphisms, but in no case has it been necessary to invoke heterozygous advantage to explain how the polymorphism is maintained. But all these examples have had one thing in common, which we have deliberately not *emphasized* until now, namely that a particular phenotype is at a relative selective advantage when *rare*. This means that a rare phenotype will increase in frequency and as it does so its selective advantage will

---

* As you will remember from earlier Units, mutations have been observed on numerous occasions in the laboratory. Indeed, selection of antibiotic-resistant individuals from a clone of susceptible bacteria is one of the simplest and yet most elegant examples both of the origin of a new form by mutation and of evolution in action.

wane. We shall now consider in more detail why frequency-dependent selection of this type will maintain a genetic polymorphism at a stable equilibrium.

You will need to be familiar with Section 9.6.1 before proceeding further.

In Section 9.6.1 we developed a general model for selection, which can be summarized as follows.

Genotype	A_1A_1	A_1A_2	A_2A_2
Genotype frequency	p^2	$2pq$	q^2
Fitness	$1 - S_1$	$1 - S_2$	$1 - S_3$

We found that the change in the frequency of A_2 from one generation to the next was given by

$$\Delta q = \frac{pq}{\bar{w}} [p(S_1 - S_2) - q(S_3 - S_2)] \tag{8}$$

where $\bar{w} = 1 - S_1 p^2 - 2Spq - S_3 q^2$. $\tag{7}$

We also found that the allele frequency at equilibrium ($\Delta q = 0$) was given by

$$q_e = \frac{S_2 - S_1}{2S_2 - S_1 - S_3} \tag{9}$$

thereby showing that the frequency at equilibrium was a function of the selective coefficients only.

We concluded Section 9.6.1 by suggesting that when the heterozygotes were at an advantage a stable equilibrium resulted. In the absence of heterozygous advantage the equilibrium was unstable and one or other of the alleles would eventually be fixed; that is, a monomorphic situation would obtain. But the fitnesses of the three genotypes were independent of their frequencies. We now need to modify our model by incorporating frequency-dependent fitnesses. We can then see whether our assertion that a polymorphism can be maintained by frequency-dependent selection will stand the test of an algebraic model, *without having to invoke heterozygous advantage*.

Let us suppose, first of all, that A_1 is dominant for fitness. If the population is a randomly mating one, A_2A_2 will have a frequency of q^2 immediately before selection. By assigning a fitness of $1 - Sq^2$ we can ensure that the fitness is a function of the frequency of this genotype. A_1A_1 and A_1A_2 will both have the dominant phenotype and their total frequency will be $1 - q^2$ immediately before selection. We can again allow for frequency-dependent selection by assigning a fitness of $1 - S(1 - q^2)$ to each of these genotypes. Our model is then:

Genotype	A_1A_1	A_1A_2	A_2A_2
Genotype frequencies	p^2	$2pq$	q^2
Fitnesses	$1 - S(1 - q^2)$	$1 - S(1 - q^2)$	$1 - Sq^2$

and is quite analogous to the model in Section 9.6.1, except that the three selective coefficients, S_1, S_2 and S_3 are now $S(1 - q^2)$, $S(1 - q^2)$ and Sq^2, respectively.

We may obtain an initial overview of the situation, firstly by making the fitnesses relative to that of the dominant phenotype (that is, the A_1A_1 and A_1A_2 genotypes). Thus

Genotype	A_1A_1	A_1A_2	A_2A_2
Relative fitness	1	1	$\dfrac{1 - Sq^2}{1 - S(1 - q^2)}$

Secondly we can plot the relative fitness of A_2A_2 against the frequency, q, of allele A_2 in the population. Let us call the relative fitness of the A_2A_2 genotype w_3 so that we can avoid having to write $(1 - Sq^2)/[1 - S(1 - q^2)]$ every time! Figure 27 shows the graphs obtained for values of $S = -1.0, 0, +1.0$. These lines intersect at $q = 0.707$. Consider first the line for $S = -1.0$. If $q < 0.707$, then $w_3 < 1$. For example, if the frequency for A_2 is 0.6, the fitness of A_2A_2 is 0.83. Because the fitness of A_1A_2 and A_1A_1 is higher (1.0), it follows that A_2A_2 will be selected against and its frequency will have fallen by the next generation. Its fitness will therefore fall again, and the process will continue leading to a decrease in q towards zero. If, however, $q > 0.707$, then $w_3 > 1$. The frequency of A_2 will then increase continually towards 1.0. In other words we have an *unstable equilibrium*. If q is moved away from 0.707

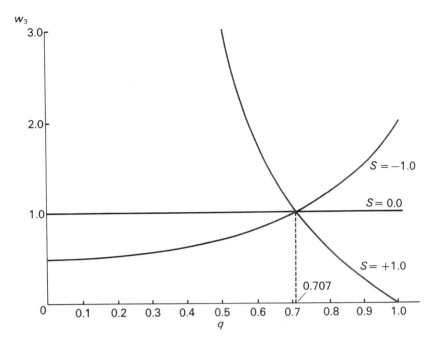

Figure 27 The relative fitness (w_3) of the recessive homozygote $A_2 A_2$ plotted against the frequency (q) of the recessive allele when subject to positive, negative and zero frequency-dependent selection.

there is no tendency for it to return to this frequency. This is true for all negative values of S, and in these cases a stable polymorphism cannot result. Examples of such 'negative frequency-dependent selection' are not entirely unknown. Thus in artificial populations of *Phlox drummondii* Levins found that plants that bore flowers with a particular heritable change (divided rather than entire petals) received fewer visits from pollinators (butterflies and moths) when plants with divided petals were scarce. If this occurs regularly in natural populations, the outcome will be the elimination of novel variants, resulting in floral uniformity.

Consider now the line for $S = 0$ in Figure 27. Whatever the frequency of A_2, the fitness of $A_2 A_2$ remains at 1, the same as that of the other two phenotypes. The net result is no change in frequency, and if q is moved away from 0.707 there is no tendency either to move further away, or to return. We have, therefore, a *neutral equilibrium*.

neutral equilibrium

ITQ 17 We have met this neutral equilibrium before. What did we call it?

Finally, when $S > 0$, for example the line for $S = +1.0$ in Figure 27, we have a particularly interesting situation. If $q < 0.707$, then $w_3 > 1$ and the frequency of $A_2 A_2$ will increase. If, however, $q > 0.707$, $w_3 < 1$ and the frequency of $A_2 A_2$ will decrease. We have, therefore, a *stable equilibrium* maintained by frequency-dependent selection. If the frequency of A_2 deviates from 0.707 it will return to this value from either direction.

stable equilibrium

ITQ 18 What are the frequencies of the dominant and recessive phenotypes at equilibrium?

We shall use our general selection model to verify this conclusion, *but you are not expected to worry about the intermediate steps in the remaining algebra of this Section.* If we substitute $S(1 - q^2)$ for S_1 and S_2, and Sq^2 for S_3 into

$$\Delta q = \frac{pq}{\overline{w}} [p(S_1 - S_2) - q(S_3 - S_2)] \qquad (8)$$

where $\overline{w}$* in this situation is given by $1 - S_1 p^2 - 2S_2 pq - S_3 q^2$ (equation 7), we obtain

$$\Delta q = - \frac{Spq^2(2q^2 - 1)}{1 - S(1 - q^2)^2 - Sq^4} \qquad (14)$$

$$\Delta q = - \frac{Spq^2(2q^2 - 1)}{1 - S(1 - q^2)^2 - Sq^4}$$

* Remember that $\overline{w}$ is the mean fitness of a population; the value of $\overline{w}$ in terms of the allele frequencies and selective coefficients will vary according to the size and type of selection being studied. Don't be worried by this: just remember that $\overline{w}$ is the mean fitness.

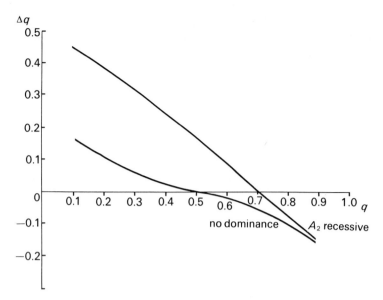

Figure 28 The change in frequency (Δq) of allele A_2 from one generation to the next plotted against its frequency (q) when subject to frequency-dependent selection.

This expression gives, as before, the change in frequency of the allele A_2 from one generation to the next. A_2 is recessive for fitness, which is also a function of the frequency of the A_2A_2 phenotype. It is an awkward formula and you do not need to remember it. We have, however, plotted a graph of Δq against q for the case where $S = 1$ (Fig. 28). We see, as we would expect, that when $q < 0.707$, Δq is positive and when $q > 0.707$, Δq is negative. Thus the frequency of A_2 always moves towards 0.707, the stable equilibrium value at which $\Delta q = 0$. This is confirmed by substituting $S(1 - q_e^2)$ for S_1 and S_2, and Sq_e^2 for S_3 into

$$q_e = \frac{S_2 - S_1}{2S_2 - S_1 - S_3} \qquad (9)$$

After rearranging we obtain

$$q_e[2S(1 - q_e^2) - S(1 - q_e^2) - Sq_e^2] = S(1 - q_e^2) - S(1 - q_e^2)$$

Therefore

$$Sq_e(1 - 2q_e^2) = 0$$

Therefore either $Sq_e = 0$ or $1 - 2q_e^2 = 0$.

But if $S \neq 0$ and $q_e \neq 0$, then

$$1 - 2q_e^2 = 0$$

Therefore

$$q_e = \sqrt{\tfrac{1}{2}} = 0.707$$

Notice that the equilibrium allele frequency is independent of the value of S, and that all three genotypes have the same fitness at equilibrium, namely $1 - S/2$. You can check this by substituting $q_e = 0.707$ into the model on p. 460. Thus, the fitness of A_1A_1 is $1 - S(1 - 0.5) = 1 - S/2$. Similarly for A_1A_2, whereas the fitness of A_2A_2 is obtained directly as $1 - 0.5S = 1 - S/2$. The heterozygotes are not, therefore, at a selective advantage. The frequency of the recessive phenotype is $q_e^2 = 0.5$ and the two phenotypes occur at equal frequency.

Let us now turn to a more general model without dominance:

Genotype	A_1A_1	A_1A_2	A_2A_2
Genotype frequencies	p^2	$2pq$	q^2
Fitnesses	$1 - Sp^2$	$1 - 2Spq$	$1 - Sq^2$

Substituting Sp^2, $2Spq$ and Sq^2 for S_1, S_2 and S_3 in equation 8 we obtain

$$\Delta q = \frac{pq[p(Sp^2 - 2Spq) - q(Sq^2 - 2Spq)]}{1 - S(p^4 + 4p^2q^2 + q^4)}$$

where

$$1 - S(p^4 + 4p^2q^2 + q^4) = \bar{w} \qquad (15)$$

$\bar{w} = 1 - S(p^4 + 4p^2q^2 + q^4)$

Again, this is an awkward expression, but in Figure 28 we have plotted Δq against q for the case where $S = 1$. Again we have a stable equilibrium point, but this time at $q = 0.5$, or $q^2 = 0.25$. At equilibrium, therefore, the genotypes A_1A_1, A_1A_2 and A_2A_2 will occur with frequencies of $\tfrac{1}{4}, \tfrac{1}{2}$ and $\tfrac{1}{4}$.

Once more we may confirm this conclusiony substituting Sp^2, $2Spq$ and Sq^2 for S_1, S_2 and S_3, respectively, in

$$q_e = \frac{S_2 - S_1}{2S_2 - S_1 - S_3} \qquad (9)$$

Or rearranging we obtain

$$q_e(4Sp_e q_e - Sp_e^2 - Sq_e^2) = 2Sp_e q_e - Sp_e^2$$

or

$$4p_e q_e^2 - p_e^2 q_e - q_e^3 - 2p_e q_e + p_e^2 = 0$$

This expression may be simplified to

$$6q_e^3 - 9q_e^2 + 5q_e - 1 = 0$$

or

$$(2q_e - 1)(3q_e^2 - 3q_e + 1) = 0$$

The only solution that gives a value for q_e between 0 and 1 is $q_e = \frac{1}{2}$, and this solution is independent of the value of S. The fitnesses of the three phenotypes at equilibrium are $1 - S/4$ for the homozygotes, and $1 - S/2$ for the heterozygotes. We have, therefore, a stable equilibrium in which the heterozygotes are actually at a selective *disadvantage* compared with the two homozygotes.

We shall now consider one particular example of frequency-dependent selection, namely apostatic selection.

10.2.1 Apostatic selection

Television programme 9 was devoted to the shell colour and banding polymorphisms of the snail *C. nemoralis*. It was suggested that a number of possible types of selection might be involved in the maintenance of these polymorphisms. One in particular was that a rare, cryptically patterned snail was at a relative selective advantage as long as it remained rare. Song thrushes develop a hunting image and tend to search for a snail similar in phenotype to the last one that they found, even if this morph is relatively infrequent and inconspicuous compared with others in the same population. Rare morphs, therefore, will enjoy a selective advantage and increase in frequency. But as they do so, their selective advantage will wane. This type of frequency-dependent selection resulting from predation is known as *apostatic selection*.

apostatic selection

A particularly convincing demonstration has come from the reanalysis of data concerning predation of the aquatic corixid bug, *Sigara distincta**, by a fish called the rudd, *Scardinius eryophthalmus*. This bug exhibits a range of brown hues, and when equal numbers of two different colours were placed on a uniform background, then those bugs that were the more cryptic suffered relatively less predation. So far there is nothing surprising. When the proportions of the two colours were varied it was found that the more cryptic bugs, although always at a relative advantage, were protected to an increasing degree as their frequency decreased. When a third, relatively uncryptic form, was offered, all three colours were found to be predated at a rate that was inversely proportional to their relative frequency. The results are shown in Figure 29. If a colour is predated at a rate consistent with its frequency in the population, the appropriate result will fall on the broken diagonal line. It is clear that even the most cryptic form was at a selective advantage when rare, but at a relative disadvantage when common.

Evidence for apostatic selection has been found in a number of studies, but perhaps the most interesting are those using artificial prey. These experiments can be conducted in any average-sized suburban garden, because the 'prey' consist of pastry 'caterpillars' dyed with edible colouring and the predators are the local population of garden birds. Thus, when brown and green pastry baits were laid out in varying proportions, it was found that blackbirds took an excess over expectation of the more common type. In other words, apostatic selection was operating. But the results were not independent of density of baits. At low density (200 baits per 100 m square) apostatic selection was found, but at high density (300 baits per 19 cm diameter circle) blackbirds took more than expected of the rarer forms. It was thought that

* *Sigara distincta* is a very small bug, closely related to the water boatman.

463

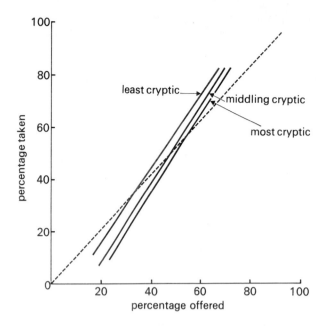

Figure 29 Frequency-dependent predation of three morphs of *Sigara distincta*, differing in their degree of crypsis, by *Scardinius eryophthalmus*.

this was because they stood out against the background provided by the more common morph. More recent work has suggested, however, that the high-density result was spurious because two trials, one of 9 brown : 1 green and the other of 1 brown : 9 green, were presented simultaneously and only 1 m apart on the ground. It is possible, therefore, that the blackbirds might have switched from one trial to the other and shown preconditioned feeding responses. This argument is supported by the results of repeat high-density experiments, but with the two trials laid 2 000 m apart. Apostatic selection was then found.

10.2.2 Summary of Section 10.2

We have shown that heterozygous advantage is by no means a necessary prerequisite for the maintenance of a stable polymorphism. Indeed, we have studied in the course of these Units a number of examples where polymorphism is maintained, in part at least, by frequency-dependent selection.

10.3 Balance between mutation and selection   [A]

Refresh your memory of the HbS *story by referring back to Section 9.4.3.*

In Section 9.5.3 we described the effects of mutation on allele frequency and argued that if we have a pair of alleles A_1 and A_2, the occurrence of mutation from A_1 to A_2 depends on the frequency of the A_1 allele and that mutation from A_2 to A_1 depends on the frequency of the A_2 allele. If A_2 is rare there will be a net excess of mutations A_1 to A_2. The increase in A_2 per generation will be proportional to the net mutation rate μ* per generation and to the frequency of A_1. Thus A_2 will increase at a rate μp per generation, where p is the frequency of A_1. The expression μp can, of course, be written as $\mu(1 - q)$.

In Section 9.6.2 we considered the special case of A_1 dominant to A_2 with the homozygote $A_2 A_2$ at a selective disadvantage (special case (a)). We saw that

$$\Delta q = -\frac{Spq^2}{1 - Sq^2} \qquad (10)$$

When the A_2 allele is rare, q is small and so q^2 is very small in relation to the 1 in the denominator. In this situation

$$\Delta q \approx -Sq^2(1 - q)$$

* Earlier we used μ for mutation of $A_1 \rightarrow A_2$ and v for mutation $A_2 \rightarrow A_1$. We now use μ for the *net* over-all mutation rate of $A_1 \rightarrow A_2$.

There can be an equilibrium when the decrease in q due to selection is exactly balanced by the increase because of mutation. Hence at equilibrium

$$Sq^2(1 - q) = \mu(1 - q)$$

that is

$$Sq^2 = \mu$$

or

$$q = \sqrt{\left(\frac{\mu}{S}\right)} \tag{16}$$

$$q = \sqrt{\left(\frac{\mu}{S}\right)}$$

Thus, if we know S or μ we can find the other in terms of q. With the example of HbS we took S as unity and so we have for that special case $q = \sqrt{\mu}$. That is, the frequency of a rare recessive lethal allele is the square root of the mutation rate to that allele.

We can now investigate the relationship between different degrees of selection against a new recessive allele arising by recurrent mutation and the equilibrium frequency of that allele in the population. Thus, if we start with $\mu = 10^{-5}$ and substitute values of S in equation 16 we obtain the results given in the following Table.

S	0.001	0.01	0.1	0.5	1.0
q^2	0.01	0.001	0.000 1	2×10^{-5}	10^{-5}
q	0.1	0.031 62	0.01	0.004 472	0.003 162

Thus mild selection ($S = 0.01$) is sufficient to keep mutant alleles at a relatively low level.

With current interest in the long-term effects of various environmental pollutants it is relevant to consider what effect a doubling of the mutation rate will have. Obviously it is not possible to perform confirmatory experiments, but we can get a good idea of the consequences from the model. Doubling the mutation rate means that μ becomes 2×10^{-5}.

ITQ 19 What effect does doubling the mutation rate have on the frequency of a disadvantageous recessive allele?

From the answer to ITQ 19 we can see that when we double the mutation rate (100 per cent increase) we obtain only a 41 per cent increase in the frequency of a new recessive allele. This looks fine for one locus and, remembering that most of these new alleles are likely to be markedly disadvantageous, we can see from the above table that the frequency of any *one* of these alleles in the population is not likely to be high. On the other hand, there is likely to be an increase in the frequency of new alleles *at all loci* and this will create problems.

In the great majority of cases the increase in the number of mutants in *haploid* organisms may not matter, because the new forms will be placed at such an enormous disadvantage in competition with the normal forms that the new allele will not survive. Even with diploids, homozygotes will occur rarely unless the population is very large.

QUESTION What is the maximum frequency that a recessive lethal allele will reach in a population of a diploid organism as a result of a mutation rate of 1×10^{-6}?

ANSWER Using the equation $q = \sqrt{(\mu/S)}$ with $\mu = 1 \times 10^{-6}$ and $S = 1$, we have $q = \sqrt{(0.000\ 001/1)} = 0.001$?

ITQ 20 Assuming that the Hardy–Weinberg law holds, what is the size of the population in the previous question that would be *expected* to produce only one recessive homozygous individual per generation?

But there are two situations in which there are clear disadvantages in any increase in mutation rate per locus. With micro-organisms there is the finite, albeit remote possibility of new pathogenic forms arising more rapidly than at present. With humans, and their domesticated plants and animals, there is the possibility of an increase both in the number of congenitally malformed offspring and in the number of individuals containing lethal mutations.

There is some direct evidence on the non-lethal effect of irradiation on humans and particularly on the difference between acute and chronic irradiation. For example, in a comprehensive and complex survey, Neel and Schull examined new-born children of 20 000 women and 14 000 men who had been sufficiently close to the atomic explosions at Hiroshima and Nagasaki to have received a dose of approximately 11 to 600 rem*.

Of 33 181 births in which at least one parent was irradiated, there were 546 still-births (1.64 per cent), whereas 31 559 births to couples, both of whom were outside these cities at the time of the explosions, contained 408 still-births (1.29 per cent).

In another series of 33 527 births in which at least one parent was irradiated, there were 300 infants with congenital malformations, whereas 31 904 births in a control group in Japan contained 294 infants with congenital malformations. Statistical analyses showed that these differences fall within sampling error and it can be argued that there was no increase in congenital defects as a result of acute irradiation.

Neel and Schull showed great ingenuity in the way they obtained their control group. In the post-war period in Japan, pregnant women who had completed the fifth month of gestation could obtain certain extra rations. When these women registered at the city halls at Hiroshima and Nagasaki they were asked, voluntarily, to fill in a questionnaire on their past reproductive history and where they and their husbands were at the time of the explosions. These data began to accumulate in 1948 and it was soon discovered that many of the women who registered were away from the two cities at the time of the explosions, but had returned or moved there subsequently. It was this group that Neel and Schull used for the controls in their survey, because it was likely that they were more similar genetically and environmentally to the exposed group than any other possible group. It also meant that the experimental and control groups could be treated in an identical manner as far as collecting and analysing the data were concerned.

In other work in the USA by Macht and Lawrence, the children of radiologists who were exposed to chronic irradiation were examined. Their results are summarized in the following Table.

	No. of births	Still-births and spontaneous abortions	Malformations
Radiologists	5 461	766	327
Non-radiologists	4 484	548	217

As far as the still-births were concerned, the results are still within sampling error, but the number of malformations is significantly higher ($P < 0.05$).

We must note, however, that:

(a) There were differences in kind between the Japanese and Americans as far as the source of irradiation was concerned. The Japanese received a single large (acute) dose, whereas the radiologists received a continuous low (chronic) dose over many years.

(b) The two studies used different criteria over choosing which defects should be included in the study.

It must be emphasized that the children were all in the first generation after irradiation. What is of paramount importance is the study of the development of these children and of the grandchildren of the irradiated people. These data are now beginning to trickle in. Later work by Neel, Schull and Kato has enabled them to calculate an estimate of the dose required to double the mutation rate in human

* These studies are plagued by complications of the units involved. The original publications measured radiation in terms of röntgens (R), which is really a unit of exposure dose to X- or γ-radiation. This unit is defined as the irradiation that produces 1.61×10^{12} ion pairs per gram of air and, in terms of energy, is equivalent to the absorption of $9.3\,\mu J$ per gram of water. But clearly this is a physical and not a biological unit. The rad is a unit of absorbed energy equal to an energy absorption of $10\,\mu J$ per gram of irradiated material and it will be different for different biological materials. The rem, which is the unit currently used, is the dose in rads multiplied by the 'relative biological effectiveness'. The rem is the preferred unit because it takes into account the differing effects of α, β, γ, X radiation and neutrons, etc. A dose of 600 rem is the maximum dose (approx.) of whole-body irradiation that an individual human can survive. (You do not need to remember these definitions.)

gametes (the so-called *doubling dose*). They have been very careful to define the terms of reference of their estimates as: 'the minimal gametic doubling dose of radiation of this type (that is, acute irradiation from nuclear explosions) for mutations resulting in death sometime during the first 17 years of life among live-born infants conceived 0–13 years after parental exposure is 46 rem for fathers and 125 rem for mothers'. They estimate the gametic doubling dose for chronic low-level irradiation as between 138 and 200 rem for males and as much as 1 000 rem for females. Their work also showed that detailed analysis failed to reveal significant variation in mortality of children ascribable to differences in the radiation received by their parents. Do note, however, that they have always been very careful not to say that the irradiation has had no effect. Even 26 years after the work began they wrote: 'As is our custom, we enter the usual caveat that this failure to demonstrate an unequivocal effect of this type does not constitute evidence [that] there were no effects!'

doubling dose

QUESTION The atomic bombs exploded at Hiroshima and Nagasaki were minute in explosive power by comparison with the atomic and hydrogen weapons of today. By comparison with the data above would you expect the use of a modern nuclear weapon to produce a higher *frequency* of still-births, spontaneous abortions and congenital malformations among the children of people who received a dose of at least 11 rem?

ANSWER Only marginally, if at all! If you answered 'yes' it is likely that you have confused the word *frequency* with *number*. Any dose above 600 rem will result in the death of a person and therefore survivors would have received a smaller dose than this. If we assume that the patterns of radiation and survival are the same as at Hiroshima and Nagasaki, we can draw the following conclusions. There would be an enormous increase in the *numbers* of people killed and in the numbers of still-born children, spontaneous abortions and congenitally deformed children produced by the survivors, but it is unlikely that there would be any increase in the *frequency* of these. The point is, that the radiation received by the survivors covers the full range from zero to lethality and there would be no increase in the frequency of people receiving higher doses, only in the numbers.

10.4 Migration

10.4.1 The effect of migration on allele frequencies

[B]

Although mutation is clearly a process of primary importance, in so far as it is the origin of genetic variation, it cannot often be a potent factor in the evolutionary process. As we have seen, the input of new alleles into a population as a result of mutation pressure alone is too slow to influence allele frequencies to any significant degree in the absence of selection (Section 9.5.3). Mutation is not, however, the only source of new genes in a population. Unless a group of individuals is reproductively isolated from other similar groups, new genes will constantly be introduced to the gene pool as a result of immigrant individuals breeding within the group.

If migrants rarely enter the population, few new alleles will be introduced. Those that are will have a status not dissimilar to that of a rare, recurrent mutant allele. But whereas the rarity of mutations is a consequence of the mutational process itself, there is no restriction on the migration rate except that imposed by the degree of isolation between two populations.

The effect of immigration on the frequency of an allele in the recipient population depends both on the migration rate and on the difference in allele frequency between the migrant and the recipient populations. The migration rate is usually expressed as the proportion of genes in the recipient population that have been incorporated as a result of migration during one generation. A simple mathematical model will make this clear.

Let us suppose that the frequency of a particular allele is q_d in the migrant (donor) population and q_0 in some initial generation in the recipient population, and that a proportion m of the genes in the recipient population enter each generation as a result of migration. We shall assume that mating is at random within the recipient population, that the allele is selectively neutral and that the population size is sufficiently large for random processes to have little influence. In the next generation a proportion mq_d of alleles will have become established in the recipient population as a result of migration.

Provided the population size remains constant a proportion mq_0 of the same allele will be lost. On balance, therefore, the allele frequency in the recipient population in this next generation will be

$$q_1 = q_0 - mq_0 + mq_d$$
$$= q_0 - m(q_0 - q_d)$$

Thus the change in allele frequency in the recipient population is given by

$$\Delta q = q_1 - q_0$$
$$= -m(q_0 - q_d)$$

and is dependent on the migration rate $(-m)$ and the difference in allele frequency between the two populations in generation 0, $(q_0 - q_d)$. The difference after one generation of migration is

$$q_1 - q_d = q_0 - mq_0 + mq_d - q_d = (1 - m)(q_0 - q_d)$$

After a further generation it will be reduced to

$$q_2 - q_d = q_1 - mq_1 + mq_d - q_d = (1 - m)(q_1 - q_d)$$
$$= (1 - m)^2(q_0 - q_d)$$

and, in general, after n generations of migration at a constant rate, m per generation

$$q_n - q_d = (1 - m)^n(q_0 - q_d) \qquad (17)$$

$$q_n - q_d = (1 - m)^n(q_0 - q_d)$$

Thus the difference between the two populations disappears at a rate $(1 - m)$ per generation. Equation 17 can be rearranged to give

$$(1 - m)^n = \frac{q_n - q_d}{q_0 - q_d}$$

Even if migration is not at a constant rate per generation, the expression

$$\frac{q_n - q_d}{q_0 - q_d} \qquad (18)$$

$$\frac{q_n - q_d}{q_0 - q_d}$$

can be used to estimate the proportion of the genes in the migrants that were derived from the ancestral population.

10.4.2 Migration in humans [B]

The Black population of the United States is largely derived from slaves who were taken from West Africa between 1619 and 1808, the great majority after 1700. The total number taken was probably somewhat less than 400 000. Soon after the slaves arrived matings between Negro women and Caucasian men began to occur, thus introducing Caucasian genes into the Negro gene pool. There was comparatively little gene flow in the opposite direction, because of the relatively small size of the Negro population, and because the progeny of these crosses were conventionally classified as 'coloured', so falling foul of the social taboos against sexual relationships between Black men and White women. In 1969 Reed applied the migration model to the allele frequencies now observed in Blacks and Caucasians in the United States and the Negroes currently living in West Africa. For example, the Fy^a allele of the Duffy blood-group system is virtually absent in West Africa, being present at a frequency probably less than 0.02. The frequency in United States Caucasians is about 0.43 and that in Blacks in Oakland, California is estimated as 0.094. Individuals were classified as Black or Caucasian according to their own opinion regarding their ethnic background. Those who considered themselves 'mixed' were not included in the study. There is no evidence for natural selection associated with Duffy blood-group phenotypes and it was suggested that the higher frequency of Fy^a alleles in United States Blacks, compared with the Africans, is due to a flow of genes from the Caucasian gene pool to the Negro gene pool.

The expression $(q_n - q_d)/(q_0 - q_d)$ was used to provide an estimate of the proportion of genes of African origin in the Californian Blacks. Here $q_n = 0.094$, $q_d = 0.43$ and $q_0 = 0.02$ (say). Thus, using expression 18

$$\frac{q_n - q_d}{q_0 - q_d} = \frac{-0.336}{-0.41} = 0.82$$

indicating an estimated 82 per cent of genes being of West African origin, and the remaining 18 per cent of Caucasian origin.

468

Furthermore, if migration can be assumed to have been maintained at a constant rate of m per generation for n generations, then

$$(1 - m)^n = 0.82$$

Taking n as 10 generations

$$\log(1 - m) = \tfrac{1}{10}\log(0.82)$$

and

$$m = 0.019\,6$$

implying that about two per cent of the genes in the Blacks are introduced from the Caucasians in each generation.

Although illustrating the principle, this answer can only be taken as very approximate. The rate of gene flow cannot have been constant over ten generations and the model does not allow for the sociocultural variables that must be associated with such a situation.

But a number of studies along these lines have now been completed in the USA and, taken *in toto*, they certainly prove most interesting.

In the northern States, for example in New York city, Detroit and Cleveland, the Black population has been found to carry a proportion of between 0.18 and 0.28 genes of Caucasian origin. In the southern States, for example South Carolina and Georgia, between 0.04 and 0.12 of the genes are estimated to be of Caucasian origin. These results are quite consistent with the more relaxed attitudes to racial integration in the northern States, although differential migration and selection cannot be ruled out. Note, however, that in studies in the southern States the loci involved have been found to fall into two quite distinct classes. The majority show the low migration rate typical of the southern States, but there are one or two loci that apparently show far higher migration rates, with up to 0.3 of the Black genes being of Caucasian origin.

Thus, for example, a study in Claxton, Georgia yielded the following results for the Hb^S allele frequencies: 0.0 for a sample of 2 090 Caucasians ($= q_d$), 0.043 for a sample of 1 287 Blacks ($= q_n$) and 0.061 1 was the average frequency of 72 West African studies ($= q_0$). These estimates yield

$$\frac{q_n - q_d}{q_0 - q_d} = \frac{0.043}{0.061\,1} = 0.703\,8$$

In other words, 0.70 of the Black genes are of West African origin and 0.30 of Caucasian origin. But this estimate of 0.30 is considerably higher than that obtained for five other loci (ABO and other blood groups) in the same samples, which yielded an average estimate of only 0.113—more typical of the southern States. It cannot be a coincidence that those loci that show the higher migration rate have alleles that, although common in West Africa, are at a distinct selective disadvantage in the southern States of the USA. Assuming that gene flow has not really proceeded at a higher rate for these loci, it is possible to estimate selective coefficients operating against these alleles by comparing the observed and expected changes in allele frequency.

QUESTION In the absence of selection against the Hb^S allele, and assuming a migration rate equal to the average of the five typical loci, what is the expected frequency of Hb^S among Claxton's Blacks?

ANSWER We need to solve

$$\frac{q_n - q_d}{q_0 - q_d} = 1.0 - 0.113$$

for q_n, where $q_d = 0.0$, $q_0 = 0.061\,1$, and $(1.0 - 0.113)$ is the proportion of negro genes among the negroes.

Therefore

$$\frac{q_n}{0.061\,1} = 0.887$$

and

$$q_n = 0.054\,2$$

There is, therefore, quite a large discrepancy between the observed frequency of 0.043 and that expected in the absence of selection. We may now return to our algebraic models of selection and attempt to determine fitnesses for the genotypes that would lead to a reduction in the frequency of Hb^S from 0.054 to 0.043 over ten generations.

If we assign fitnesses of

Genotype	Hb^AHb^A	Hb^AHb^S	Hb^SHb^S
Fitness	1	1	0

(in other words Hb^S is a recessive lethal, and there is no heterozygous advantage in the absence of malaria), we may apply equation 4

$$q_n = \frac{q_0}{1 + nq_0}$$

Now since $q_0 = 0.054$ and $n = 10$ generations, we find that

$$q_n = \frac{0.054}{1.54} = 0.035$$

We *observed* a reduction to 0.043, and our model has given us a reduction to 0.035.

The discrepancy is not large, given the considerable number of assumptions and estimates used in our models. None the less, the expected reduction in frequency is of an order of twice the magnitude of the observed reduction. Relaxation of selection operating against the Hb^SHb^S homozygotes will reduce the expected reduction, but repeated trial-and-error application of the appropriate equation

$$\Delta q = -\frac{Spq^2}{1 - Sq^2} \tag{10}$$

only yields a realistic result once the fitness of the Hb^SHb^S homozygotes has reached 0.5. These individuals probably do have a fitness somewhat higher than zero, say 0.2, but certainly not as high as 0.5. If we care to invoke heterozygous advantage the appropriate equation is

$$\Delta q = \frac{pq(S_1 p - S_3 q)}{1 - S_1 p^2 - S_3 q^2} \tag{13}$$

and the required change may be obtained in 10 generations with, for example, the following combination of fitnesses

Genotype	Hb^AHb^A	Hb^AHb^S	Hb^SHb^S
Fitness	$1 - S_1$	1	$1 - S_3$
	0.97	1	0
	0.985	1	0.2

These slight levels of heterozygous advantage are not inconsistent with the former presence of falciparum (subtertian) malaria in the southern United States. The model is, however, still naïve because it assumes that the relative fitnesses of the three genotypes have remained constant over all ten generations. This clearly cannot be the case.

We have now gone a long way towards answering the question originally posed in Section 9.4.1.

Note that the estimates we have considered in this Section have referred to populations and *not* to individuals. It has recently been suggested, however, that this method could be used to estimate the degree of hybridity of individuals. Quite apart from the questionable desirability of such an undertaking, Reed has shown that it is not feasible on purely genetic grounds. In order to establish the proportion of Caucasian genes in a Negro individual with a 95 per cent confidence interval of width 0.1* (see STATS, Section ST.8 if you are not sure about confidence intervals), we would require a minimum of 72 *ideal* marker loci. An ideal locus is one for which there is a large difference in allele frequencies between the Caucasian and West African populations, as was the case with Duffy allele Fy^a. The lower the difference, the greater is the variance of the estimate. At present we do not have 72 loci with alleles present at high frequency in American Caucasians, but missing in West Africans, or vice versa. Furthermore, to be ideal loci, all 72 must be unlinked. Humans have only 23 linkage groups!

* By 'width 0.1', we mean that if the estimate is 0.45, we can be 95 per cent certain that the true value lies between the limits 0.5 and 0.4.

10.4.3 Migration versus mutation

Outside the realms of human politics and the imposition of immigration and emigration quotas, a steady-state system of migration is not easy to imagine for any other organism. The theoretical considerations of a constant migration rate give rise to models that parallel those of mutation, and the conclusions are, essentially, the same. If migration is extensive and selection is weak, then the sub-populations that share the migrants will tend to become alike. On the other hand, if the selection in the different sub-populations is much larger than the migration rate, then there will be considerable local differentiation of the sub-populations.

Occasionally an elegant experiment is devised that not only answers one very specific question, but also has ramifications into other problems. Such a one was reported briefly by Sokal to the XIIIth International Congress of Genetics in 1973. Sokal showed that a newly arisen mutant allele stands a greater chance of survival in a population than the identical allele arriving in that same population by migration. Read the edited abstract* below and you will then be in a position to answer the question following.

> Two strains of flour beetles were used for this work, UPF and Chicago Black. Forty population cages, each with 500 adult *T. castaneum* of the UPF strain, received a *bb* (black) female newly mated with a UPF male. Half of the immigrants had a Chicago Black genetic background, the other half a UPF background. These conditions simulate, respectively, the fate of a rare, genetically-differing immigrant or the fate of a mutation in populations of considerable size. Adults were censused for 11 discrete generations... The semi-dominant black allele survived in 26 out of 40 cultures by the end of the experiment, demonstrating its selective advantage at these very low frequencies. The allele increased from an initial frequency of 0.002 to 0.055 (at generation 11) in at least one replicate. Mean frequencies at generation 11 were 0.053 (alien background) and 0.066 (native background). The cultures simulating mutations (immigrants with native backgrounds) had a higher average allele frequency, different distribution of allele frequencies across replicates, and a lower extinction rate of black than the cultures with alien background immigrants.

> QUESTION Suggest a reason for the different results from the two parts of the experiment.

> ANSWER Even at these low frequencies the genetic background of the immigrants plays an important role in determining the outcome of the experiment.

This experiment by Sokal shows another of the weaknesses of the theoretical approach. In order to develop the elementary models we have had to make some basic assumptions, but not infrequently we do not know what these assumptions are. It is often only after we have obtained data that are not explicable by the model that we begin to question some of the premises. It should have been obvious that the genetic background in which a new mutant arises should have some influence on the survival of that allele, but until Sokal actually showed that the genetic background could be important it was not afforded much credence.

10.5 Small populations and random effects

We have seen how mutation, migration and selection can change the frequencies of alleles in populations. The important feature common to all these processes is that the magnitude and direction of the change can be predicted, given that we know the allele frequency and also the mutation rate, migration rate or selective differential, as appropriate. Furthermore, these processes operate in such a way as to lead to the eventual fixation, or loss, of an allele unless they are opposed in their direction of influence, in which case a stable equilibrium will result. For this reason, mutation, migration and selection may be regarded as being *systematic* in their mode of operation.

In practice, however, a predicted change in allele frequency will rarely be achieved. Consider a mutation rate from A_1 to A_2 alleles of 10^{-5} per generation. This implies that one in every 100 000 A_1 alleles might be expected to mutate to A_2 during each generation. The operative word here is 'expected'. It would be surprising if *exactly*

* From Sokal, R. R. (1973). The fate of single mutant immigrants in large *Tribolium* (flour beetle) populations. *Genetics, Princeton*, **74**, 260–1s.

one in every 100 000 A_1 alleles mutated in every generation. By the very nature of the variability inherent in biological systems, the number of mutations observed in some generations should be somewhat less, and in others somewhat more, than that expected. Greater deviations from expectation should occur less often than small deviations and, over a relatively large number of generations, the average number of mutations should agree well with the mutation rate. Similar considerations also apply to selection and migration. Although all three systematic processes are directional and influence allele frequencies in a predictable manner, their operation is subject to random fluctuations.

Consider now a population in which mutation, migration and selection do not operate. In such a population, alleles are of neutral selective value with respect to each other, or, more generally, they are *neutral alleles*. The Hardy–Weinberg law predicts that the allele frequencies and, if mating is at random, the genotype frequencies should remain constant from generation to generation. In Section 9.4.4 we listed the assumptions necessary for this law to hold and these included a large population size and no effects resulting from random processes. Throughout these Units we have, in fact, been making these two assumptions and we shall see in the following Sections that a small population size leads to random fluctuations in allele frequency.

neutral alleles

10.5.1 Random genetic drift

Suppose we have a population of marine hermaphrodites that maintains a constant size of only five individuals. Each member of the population simply sheds large numbers of male and female gametes into the surrounding water, and fertilization occurs between random pairs of gametes. Let us further suppose that at a particular locus, alleles A_1 and A_2 occur with frequencies p_0 and q_0 among these five individuals in generation 0. The two alleles will also be represented among the gametes at these same frequencies.

QUESTION If the allele frequencies are equal in the parental population (that is, $p_0 = q_0 = 0.5$), what is the probability that, among the progeny, A_1 will occur with a frequency $p_1 = 0.5$, or $p_1 \neq 0.5$ in the next generation? You may find it helpful to refer to *STATS*, Section ST.3.

ANSWER There will be $N = 5$ progeny in generation 1 because the population size remains constant. Therefore, from among the pool of gametes, half of which carry A_1, we require the probability that of the $2N$, or 10, gametes that contribute to the next generation, five will carry A_1. This probability is found from the appropriate term in the binomial expansion of $(p_0 + q_0)^{2N}$, which is

$$\frac{(2N)!}{5!(2N-5)!}(p_0)^5(q_0)^{(2N-5)} = \frac{10!}{5!5!}\left(\frac{1}{2}\right)^{10} \approx 0.25$$

or one-quarter. It follows that the probability that the allele frequency equals some value other than 0.5 in generation 1 is given by $1 - 0.25$, or three-quarters.

This result demonstrates the importance of population size. There is no migration, mutation or selection operating in this population, mating is at random, and yet it is very likely that there will be a change in allele frequency between one generation and the next. All the restrictions necessary for the Hardy–Weinberg law to hold are being applied, except for the fact that the population size is small.

This situation is exactly analogous to a tossing experiment with an unbiased coin. The principle involved is explained in *STATS*, Section ST.3, p. 17, where we discuss the binomial distribution. Out of any sample of throws, the observed proportion of heads may be greater than, or less than, one half by any amount, yet every possible outcome has a defined statistical probability of occurring. As the sample size, or the number of throws, increases, the probability of obtaining equal *numbers* of heads and tails becomes smaller, and yet the *proportions* of heads and tails approximate more closely to one-half each. In our present genetic example, each gamete contributing to the next generation is equivalent to the throw of a coin, and the two different alleles are equivalent to heads or tails. The number of gametes (twice the population size) required to form the next generation is analogous to the number of throws in a sample. The allele frequency may be increased or decreased as a result of this random sampling of gametes, but the expected allele frequency in the next generation is the same as the frequency in the initial generation. This expected frequency is the same as that obtained by averaging over a large number of similar small populations, or the same as that obtained in a single very large population.

In summary, therefore, random sampling of gametes in the formation of each new generation causes changes in allele frequency in one or other direction in each successive generation without any predictable constancy. It is a dispersive process and cannot by itself cause the allele frequencies to reach an equilibrium. This is known as random genetic drift and the magnitude of its effect increases with decreasing population size.

10.5.2 The cumulative nature of drift [B]

What is the effect of drift over a number of generations? In small populations the allele frequencies will fluctuate from generation to generation because each new generation's gene pool is a small sample taken at random from the parental gene pool, and is therefore subject to sampling variation. We can imagine a large number of small populations all with N members, formed by the subdivision of a single large interbreeding group. Let alleles A_1 and A_2 have initial frequencies p_0 and q_0 in each of these populations. In the next generation the frequency of A_1 in the populations is expected to follow a binomial distribution with a mean of $\bar{p}_1 = p_0$ and variance of $V(p_1) = p_0(1 - p_0)/2N$. Note that because we are finding the mean and variance of allele frequencies the appropriate expressions are those for a binomial distribution of proportions (see $STATS$, Section ST.3).

Note also that the term for the population size, N, occurs in the denominator of the variance, the effect of drift will become progressively less as the population size increases.

For example, if $p_0 = 0.5$ and $N = 5$, then (see $STATS$, Section ST.12.3)

$$V(p_1) = \frac{p_0(1 - p_0)}{2N}$$

$$= \frac{0.5 \times 0.5}{10} = 0.025$$

and the standard deviation

$$SD(p_1) = \sqrt{0.025} = 0.158$$

By approximating to the normal curve (which is *very* approximate when the population size is less than 15) this implies that:

68 per cent of populations have an allele frequency in the range 0.5 ± 0.16

95 per cent of populations have an allele frequency in the range 0.5 ± 0.31

99 per cent of populations have an allele frequency in the range 0.5 ± 0.41

(see $STATS$, Section ST.6.1, p. 31). It equally implies that the change in allele frequency is:

less than 0.16 in 68 per cent of populations

less than 0.31 in 95 per cent of populations

less than 0.41 in 99 per cent of populations.

This is shown diagrammatically in Figure 30 for 95 per cent of the populations.

For comparison, if $p_0 = 0.5$ and $N = 50$, then $V(p_1) = 0.002\,5$ and $SD(p_1) = 0.05$. Figure 31 shows the distribution of allele frequencies, again for 95 per cent of the populations.

This demonstrates very clearly that the effect of random drift decreases as the population size increases.

In the first generation, therefore, the allele frequencies in the populations will have dispersed, the extent depending on the population size. Sampling errors in the formation of generation 2 will lead to further spreading of the allele frequencies. But this time each population will start from allele frequencies different from those in

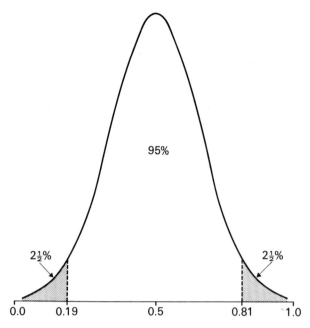

Figure 30 Distribution of allele frequencies in populations of size five after one generation of drift, with an initial allele frequency of 0.5.

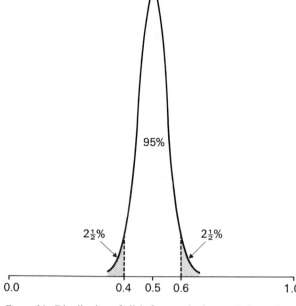

Figure 31 Distribution of allele frequencies in populations of size 50 after one generation of drift, with an initial allele frequency of 0.5.

generation 1. Therefore, the variance of the change now differs between the populations because it depends on p_1, the allele frequency in the previous generation in that population. The algebra now gets rather more difficult, but it may be shown that the average, or expected, allele frequency over all populations is

$$\bar{p}_2 = p_0$$

and the variance in allele frequency between all populations is

$$V(p_2) = p_0(1 - p_0)\left[1 - \left(1 - \frac{1}{2N}\right)^2\right]$$

In general, after n generations

$$\bar{p}_n = p_0$$

and

$$V(p_n) = p_0(1 - p_0)\left[1 - \left(1 - \frac{1}{2N}\right)^n\right]$$

Now, as n increases, $(1 - 1/2N)^n$ decreases, and the variance becomes larger with each successive generation. This means that as far as their allele frequencies at this locus are concerned, the populations are diverging. After a large number of generations, although the mean remains unchanged, the variance increases with time and, in the limit, tends to $p_0(1 - p_0)$. This implies that eventually there will be a proportion, p_0, of populations in which allele A_1 is fixed (that is, its frequency = 1.0) and in the remaining proportion, $q_0 = (1 - p_0)$, the allele A_2 is fixed and A_1 lost. For example, if there were initially 100 populations with $p_0 = 0.5$, after a very long time it is to be expected that A_1 will be fixed in 50 of these populations, but lost in the other 50. If at the start, however, the allele frequency was 0.1, then eventually A_1 can be expected to be fixed in only 10 out of the 100 populations.

This argument leads to an important result. At the very end of the process A_1 has a probability of fixation equal to its frequency at the start. Therefore, if a selectively neutral allele arises by mutation just once in a population of size N, the chance that this allele is ultimately fixed is $1/(2N)$. This probability becomes smaller as the population size increases, tending to zero as the population size tends to infinity.

As drift proceeds, the probability that both alleles coexist within a population will be continually reduced. Concomitant with this must be a reduction in the frequency of heterozygotes, generation by generation, within each population. In other words, we may regard the fixation or loss of an allele as the loss of a heterozygote and the mean time (number of generations) required for this to occur can be shown to be (take the equation for granted)

$$\bar{n} = -4N[p_0 \log_e p_0 + (1 - p_0)\log_e(1 - p_0)]$$

474

This is at a maximum after $2.8N$ generations, when $p_0 = 0.5$. The mean fixation time is shorter when p_0 is near 0 or 1.0, as is to be expected. For example, if $p_0 = 0.05$, the mean fixation time is $0.8N$ generations. These represent extremely long periods of time if the population size is anything but very small.

We have been considering the fate of alleles at a single locus in a number of separate populations. The same conclusions apply equally to the fate, within a single isolated small population, of alleles of a number of separate loci, all with the same initial frequencies.

We may now summarize the three main effects of random genetic drift on the frequency of neutral alleles in small populations. It is assumed that there is no selection, mutation or migration.

(a) There is differentiation between the allele frequencies in different populations. They increase in some, and decrease in others. Eventually this leads to random loss and fixation of alleles.

(b) There is a reduction in the genetic variation within a population, but the variance of allele frequencies between populations increases.

(c) There is a decrease in the frequency of heterozygotes.

10.5.3 Random genetic drift in natural populations

Drift will undoubtedly influence allele frequencies in small laboratory populations and in small inbred herds or flocks. Even in large populations, drift can easily cause alleles with frequencies near 0 or 1 to become lost or fixed as the result of a chance fluctuation. But in large populations new mutations soon re-introduce lost alleles. Whether or not drift can cause random fluctuations in allele frequencies in natural populations to a sufficient degree to be regarded as having any role in evolution, is another matter. What is not in question is the mathematical theory, which is generally accepted as being correct. The main problem is whether or not the dispersive nature of drift can be maintained in the face of the operation of systematic evolutionary forces, in particular selection. The models we have used so far in our discussion of drift have assumed small populations, and also negligible mutation, migration and selection. This implies that alleles are selectively neutral with respect to one another, and it is seriously doubted by some that truly neutral alleles exist in nature. These ideas will be developed further in Section 10.6.

It must be stressed that population size is not the only factor to be considered when assessing the role of drift in evolution. The extent to which drift may occur will also depend on the magnitude and direction of mutation, migration and selection pressures. Thus drift will have a greater influence on the frequency of selectively neutral alleles in relatively large populations than on the frequency of alleles subject to large selective pressures in considerably smaller populations.

Considering mutation, migration and selection separately, we may consider, as a rule of thumb, a population to be small, in the sense that drift will predominate, if $4N\mu$, $4Nm$ or $4NS$ are less than 1. On the other hand, if one of these expressions is greater than 1, the population will behave as if it were large and drift will be swamped by the effects of mutation, migration or selection. Thus, if a population numbers 2 500 individuals, a selective coefficient as small as 0.000 1 is sufficient for changes in allele frequency to reflect the operation of selection. If a selective coefficient is larger, say 0.01, then the effects of selection will predominate in even smaller populations, in this case one of only 25 individuals.

10.5.4 The founder principle

When a new population is initiated by a small number of migrants, or in an extreme case, by a single fertilized female, it is extremely unlikely that the founders will be genetically representative of the population from which they came. In such circumstances, whatever genes the founders take with them, whether or not they are advantageous, stand a chance of becoming established in the early stages of the new population. Mayr has called this the *founder principle*. It will lead to populations derived from different founders being genetically distinct from one another and also from the parent population. This may be regarded as a special case of genetic drift. Random factors operate at the time of foundation, but once a new population increases in size, selection will operate on whatever genetic variability there is within a

founder principle

particular population. The evolutionary outcome is that populations will differentiate, not because of the continuous operation of drift, but through the influence of selection on varying initial genetic inputs into the different populations.

The founder principle has been experimentally demonstrated by Dobzhansky and Pavlovsky (1957) in laboratory populations of *D. pseudoobscura*. The parent population consisted of F_2 hybrids between a stock from California with the gene arrangement Arrowhead (AR) on the third chromosome and a stock from Texas with the sequence Pikes Peak (PP) (see Section 10.1.4). Thus the two types of chromosome were expected to be present at equal frequency in the parent population. 4 000 founders were taken at random for each of ten 'large' populations, and 20 founders for each of ten 'small' populations. Because of the high fertility of *Drosophila* this initial difference in population size was rapidly lost, and all population cages soon equilibrated to roughly equal population densities. Figure 32 shows that after 17 months the frequency of the PP chromosome arrangement varied between 20 and 35 per cent among the ten 'large' populations. The variation in frequency among the ten small populations was significantly higher, however, ranging from 16 to 47 per cent.

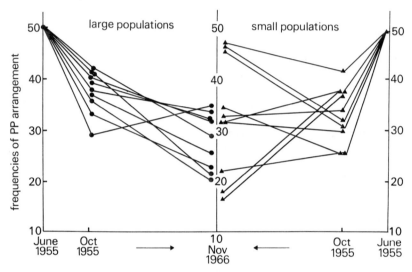

Figure 32 Frequencies of the PP chromosome arrangement in population cages initiated either by 4 000 founders (large populations) or by 20 founders (small populations).

ITQ 21 Interpret the results of this experiment.

10.5.5 Effective population size

We have shown that unless a population is infinite in size genetic drift will cause allele frequencies to fluctuate at random. The magnitude of this effect is inversely proportional to the size of the population. We have, however, been assuming non-overlapping generations, a constant population size, and that every individual contributes equally to the next generation. To elaborate our simple models to allow for these complications would introduce algebraic problems of considerable complexity.

But there is a way of avoiding these problems. Let us call the population size that we obtain by counting heads, the *census* population size. This is the actual number of individuals in the population at a given time and we have so far employed the symbol 'N' for this quantity. We now argue as follows. Can we define a population size, called the *effective population size*, N_e, that will give the same amount of drift in a simple model, such as those we have used so far, as would be obtained from a more complicated model that takes account of fluctuating population sizes, different contributions from the two sexes and overlapping generations?

effective population size

The answer is yes, and the effective population size is always less than the census population size. The mathematical arguments are complicated, and we shall merely quote the results.

Changes in population size

In 1931 Wright showed that if the census population size changes with time, then the effective population size can be calculated from the population sizes in each successive generation; that is

$$\frac{1}{N_e} = \frac{1}{n}\left[\frac{1}{N_1} + \frac{1}{N_2} + \cdots + \frac{1}{N_n}\right]$$

For example, if the population size takes values $10, 10^2, 10^3, \ldots, 10^6$, in six successive generations, then the effective population size turns out to be 54 individuals. Thus, although the population size increases from ten to one million in six generations, over all it behaves as one with a constant size of only 54 individuals. This surprising result shows that the periods when a population is small have a disproportionate effect.

Different contributions from the two sexes

When there are unequal contributions from the two sexes, the effective population size for different numbers of males (N_m) and females (N_f) is

$$N_e = \frac{4N_m N_f}{N_m + N_f}$$

Non-reproducing individuals should not be included in N_m or N_f.

Polygamy, as practised in some human societies, will cause a reduction in the effective population size. With an average of two women per married man, as is found in some African tribes, we find that the reduction is only small. Put $N_f = 2N_m$. Then

$$N_e = \frac{8N_m^2}{3N_m} = 2.67N_m$$

whereas $N \approx 3N_m$.

But if the numbers of reproducing males and females differ greatly, then the effective population size depends mainly on the less numerous sex. For example, consider a herd, or harem, with one male and 100 females.

$$N_e = \frac{400}{101} \approx 4$$

and we obtain approximately this answer whenever one sex greatly predominates. Thus, although this herd contains 101 individuals, as far as the effect of drift on allele frequencies is concerned, it will behave like a population containing only four individuals!

Overlapping generations

The problem of overlapping generations has already been introduced in Section 9.4.4 when we were discussing the assumptions on which Hardy–Weinberg equilibrium is based. At any one time only a proportion of the total number of individuals in the population will be of reproductive age. In humans this proportion is about one-third. This means that when generations overlap we cannot take the census population size as being equal to the effective population size. One suggested correction is as follows. Put

$$N_e = tN^*$$

where t is the mean age at reproduction, measured in some convenient time unit, and N^* is the number of individuals who reach this mean reproductive age during one time unit. As an example, trial figures for a group of 1 000 humans might be $t = 25$ years, and $N^* = 20$ individuals per year. Then the effective population size is only 500.

If the contributions of the two sexes differ, this formula can be applied separately to each sex and the answers entered into

$$N_e = \frac{4N_m N_f}{N_m + N_f}$$

You should now attempt SAQ 10.

10.6   Variation and the rate of evolution

We now move into a field in which there has been great controversy for several years. By the time we reach the end of this Section there will be few conclusions that we shall be able to draw other than that much work remains to be done. It is here, perhaps, that the conflict between theoretical genetics and what happens in natural populations is the most fierce. Both sides of the argument are probably correct under many circumstances, but we have yet to clarify what these circumstances are.

10.6.1   The rate of evolution   [B]

A problem about which models can tell us *something* is the change in the frequency of alleles under selection depending on their dominance relationships and their rarity. Let us return to our first special case in Section 9.6.2, which subsequently we showed (at the beginning of Section 10.3.1) to be $\Delta q \approx -Sq^2(1-q)$ when the A_2 allele is rare. This equation takes a negative attitude to the problem because it considers selection against the $A_2 A_2$ homozygote, that is, the maintenance of a *status quo*. Instead, let us now invert the situation, taking a positive attitude and think of selection favouring the A_1 allele. Fortunately this makes very little difference to the equation at this stage: it becomes

$$\Delta p \approx Spq^2$$

or

$$\Delta p \approx Sp(1-p)^2$$

Suppose that $S = 0.2$ and we substitute different values for p between 0 and 1. We can now plot values of Δp against the corresponding values of p on a graph. By inspection of the graph in Figure 33 we can see that the most rapid changes in allele frequency—that is, the largest values of Δp—occur in the intermediate range of values of p. Using this model, we can make some predictions. When an allele is rare, selection in favour of a new dominant allele increases its frequency more rapidly than will the same selection acting in favour of a rare recessive. The reason for this should now be clear; a recessive allele is nearly always heterozygous when rare ($2pq \gg q^2$) and so it is very rarely exposed to selection.

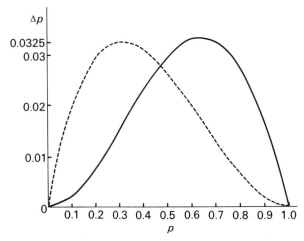

Figure 33 Graph of Δp plotted against p, allowing for different dominance relationships between the A_1 and A_2 alleles, but with the phenotype determined by $A_1 A_1$ having a selective advantage.

- - - - A_1 dominant and advantageous; calculated from $\Delta p = \dfrac{Spq^2}{1 - Sq^2}$

———— A_1 recessive and advantageous; calculated from $\Delta p = \dfrac{Sp^2 q}{1 - S(2pq + q^2)}$

We can also use the basic models to determine the change in the frequency of alleles over several generations. If Δp is the change in the value of p and the rate of change is slow, then the change in p with time is approximately dp/dn.

Taking $dp/dn = Sp(1-p)^2$, we can solve the differential equation to obtain the values of n (in generations) that are required for specific changes in the frequency of q. We could also allow for dominance by using the first three of our special cases in Section 9.6.2. When we integrate between the limits p_0 and p_n (the values of p for the initial generation and generation n) and rearrange the resulting equation in terms of n, we obtain for an advantageous dominant allele

$$n = \frac{1}{S}\left\{\log_e\left[\frac{p_n(1-p_0)}{p_0(1-p_n)}\right] + \frac{1}{1-p_n} - \frac{1}{1-p_0}\right\}$$

for an advantageous recessive allele

$$n = \frac{1}{S}\left\{\log_e\left[\frac{p_n(1-p_0)}{p_0(1-p_n)}\right] - \frac{1}{p_n} + \frac{1}{p_0}\right\}$$

and for a heterozygote of intermediate fitness with no dominance (that is, where the relationship between S_1, S_2 and S_3 is $S_1 = 0$, $S_2 = S/2$ and $S_3 = S$)

$$n = \frac{2}{S}\log_e\left[\frac{p_n(1-p_0)}{p_0(1-p_n)}\right]$$

Remember that p designates the frequency of the advantageous allele and n is the number of generations required to change the frequency from p_0 to p_n.

QUESTION What is the relationship between the time taken for a given change in allele frequency and the selective coefficient?

ANSWER They are inversely proportional to each other; that is, the greater the difference in fitness between the phenotypes, the more rapid is the change in allele frequency. This is, of course, exactly what we would expect.

Using these equations it is not difficult to calculate the number of generations required for a specified change in allele frequency for various selective coefficients.

Thus when $S = 0.01$ we obtain the results summarized in Table 21*.

Table 21 Number of generations required to effect changes in allele frequency

Frequency p	Number of generations to effect the change in allele frequency		
	$h = 0$ A_1 dominant	$h = 1$ A_1 recessive	$h = \frac{1}{2}$ heterozygote of intermediate fitness
0.000 01 → 0.01	693	9 990 692	1 384
0.01 → 0.1	250	9 240	480
0.1 → 0.5	308	1 019	439
0.5 → 0.99	10 259	558	919
0.99 → 0.999 99	9 990 692	693	1 384
0.5 → 0.858 579	687		

The effect on the number of generations of different selective coefficients is easily obtained by multiplying the entries in the table by $0.01/S$. Thus if S is 0.000 1 each of the entries needs to be multiplied by 100; that is, it would take one hundred times as many generations at each step in the series listed.

QUESTION Suggest which of the assumptions implicit in these equations are unlikely to be true in any real case.

ANSWER (a) It is highly unlikely that the selective coefficient remains constant over the whole range of allele frequencies.

(b) It is certainly unreasonable to expect the selective coefficient to remain constant over the ten million generations (approx.) that it takes for an advantageous recessive allele to increase in frequency from 0.000 01 to 0.01!

(c) When an allele is rare it is quite likely to be lost because by chance it does not appear in any individual of the next breeding generation.

* See p. 421 if you have forgotten the meaning of the symbol h.

All these points have been discussed earlier in these Units but it is only the last one that we shall consider in more detail now.

Although the question just posed reveals serious limitations to the equations on p. 479, they do give us a general indication of the qualitative effects of dominance and recessivity on the times involved in evolutionary change, and they can be used to estimate, very roughly, the selective pressure required to change, for example, the frequency of the *carbonaria* form in *B. betularia* during the past 100 years or so. Among the assumptions we have to make is that the *carbonaria* homozygote is as fit as the heterozygote (which it is not!), the frequency of the *carbonaria* allele at the beginning was 0.000 01 (which is probably too high!) and the selective coefficient is the same throughout the time involved (which we have already said is unlikely!). Notice that the numbers of generations in Table 21 can be added. That is, it takes $250 + 308 = 558$ generations to change the frequency of an advantageous dominant allele from 0.01 to 0.5.

QUESTION Assume that the frequency of the *carbonaria* allele when it was first recorded in Manchester in 1848 was 0.000 01 and that the beast goes through one generation each year. By 1900 the black form was prevalent, although the typical still occurred. In 1952 the frequency of the *carbonaria* form was 98 per cent. Calculate the average selective coefficient required to change the frequency of the *carbonaria* allele from 0.000 1 in 1848 to what it was in 1952. There is an extra entry in Table 21 to help you.

ANSWER The first thing we must do is to estimate the frequency of the *carbonaria* allele in 1952.

CC	Cc	cc	Total
98		2	100

The frequency of the C allele is $1 - \sqrt{(2/100)} = 0.858\ 579$.

Therefore

$$p_0 = 0.000\ 01$$

and

$$p_n = 0.858\ 579$$

From Table 21 we see that the number of generations required to bring about this change with a selective coefficient of 0.01 is

$$693 + 250 + 308 + 687 = 1\ 938$$

The time period involved was 104 years and so we can assume that there were only 104 generations available to effect the change. Thus we can write

$$\frac{0.01}{S} \times 1\ 938 = 104$$

Therefore

$$S = \frac{0.01 \times 1\ 938}{104} = 0.186$$

So, without placing too much confidence in the estimate, we can conclude that the selective advantage of the *carbonaria* allele over the total period for 1848 to 1952 was approximately 19 per cent. This is a high selective advantage, but even more powerful selection is implied by the hearsay evidence that the frequency of the *carbonaria* form in Manchester had already reached 98 per cent by 1895. In 1924 Haldane estimated—and you have the means of checking this—that on average the *carbonaria* form had an advantage of about 50 per cent over the typical form during the period 1848 to 1895.

When Fisher wrote the first book that linked together Mendelian, quantitative and population genetics (*The Genetical Theory of Natural Selection* (1930)) he suggested selective advantages of up to 0.01 in wild populations. We now know that selective coefficients of 0.2 to 0.3 are by no means uncommon in the species that have been studied since 1930. But it needs to be remembered that people have deliberately chosen to study species in which significant changes in frequency have been detected. You now have enough information available to calculate the selection required to bring about a significant change in allele frequency in a specified time in a population of a specific size.

It would not be fair to ask you to answer the question below either as an ITQ or an SAQ, but we shall work through it because it shows the use that can be made of the models when examining wild populations of animals or plants.

QUESTION Suppose we are studying a polymorphic population of a diploid organism that reproduces only sexually, has one generation per year and survives the winter in the egg. Each year we score 1 000 individuals for the two phenotypes determined by the allele pair A and a (that is, AA and Aa individuals have the same phenotype). What is the minimum selective advantage that can be detected by a significant change in phenotype frequencies over a period of four generations? Assume that the two phenotypes occur at equal frequency in the first sample and that selection favours the dominant phenotype.

ANSWER First of all we must calculate the minimum change in allele frequency that will give a significant value for χ^2 at the 5 per cent level. The smallest divergence from 500 : 500 that will give a significant value for $\chi^2_{[1]}$ (that is, $\chi^2_{[1]} \geqslant 3.84$) can, for the purpose of this question, be calculated on a trial-and-error basis. In fact, it is

$$
\begin{array}{cc|c}
500 & 500 & 1\,000 \\
544 & 456 & 1\,000 \\
\hline
1\,044 & 956 & 2\,000 \quad \chi^2_{[1]} = 3.88
\end{array}
$$

Sample of the zeroth generation

AA	Aa	aa	Total
500		500	1 000

$$\text{frequency of } A = 1 - \sqrt{\left(\frac{500}{1\,000}\right)}$$

$$= 0.293$$

Sample of the fourth generation

AA	Aa	aa	Total
544		456	1 000

$$\text{frequency of } A = 1 - \sqrt{\left(\frac{456}{1\,000}\right)}$$

$$= 0.325$$

Therefore

$$p_n = 0.325, \quad 1 - p_n = 0.675$$
$$p_0 = 0.293 \quad \text{and} \quad 1 - p_0 = 0.707$$

Substituting these values in the equation for selection in favour of a dominant allele

$$n = \frac{1}{S}\left\{\log_e\left[\frac{p_n(1 - p_0)}{p_0(1 - p_n)}\right] + \frac{1}{1 - p_n} - \frac{1}{1 - p_0}\right\}$$

we have

$$4 = \frac{1}{S}\left[\log_e\left(\frac{0.325 \times 0.675}{0.293 \times 0.707}\right) + \frac{1}{0.675} - \frac{1}{0.707}\right]$$

$$S = \frac{1}{4}\left[\log_e 1.059 + 1.481 - 1.414\right]$$

$$S = 0.25(0.057 + 0.067)$$

$$S = 0.25 \times 0.124$$

$$S = 0.031$$

481

An exercise like this shows how difficult it is to detect selection acting in the wild. In this example it required a selective coefficient of over 3 per cent to change the frequency of the phenotypes to an extent that could give rise to a significant value of χ^2.

10.6.2 Evolution at the molecular level

In Section 10.1.1 we described some of the work of Horowitz on the tyrosinase enzyme of *N. crassa* and stated that further discussion would be left until later. Now is the time for us to consider, in more detail, the methods of detection and the significance in evolution of changes in the structure of single proteins.

Proteins can in a sense give almost direct chemical evidence of genetic change.

QUESTION What do we mean by the above statement?

ANSWER The amino acid sequence in any particular protein depends on the sequence of base pairs in a specific gene. Consequently where changes in amino acid sequence in proteins can be detected they will reflect corresponding changes in the gene.

In this Section, we shall not become involved in the intricacies of amino acid sequencing and so your knowledge of the primary, secondary and tertiary structures of proteins from S100 will be sufficient. What is important, however, is to appreciate the role of the information gained by such techniques, in our general understanding of evolution.

There are three major areas where the study of protein chemistry has helped, and is helping, to elucidate the complexity of evolution, but we have room to consider only two of them in these Units. Firstly, there is the construction of evolutionary relationships between related individuals based on the similarity in amino acid sequence in various proteins, for example haemoglobin and cytochrome c (this Section and Television programme 14, Molecular Evolution). Secondly, investigations into altered amino acid sequences in individual proteins have been related to their morphological and physiological manifestations, for example, molecular diseases like sickle-cell anaemia. (See, for example, Neel, J. V. (1974). 'Inferences concerning evolutionary forces from genetic data.' In *Genetic Polymorphisms and Diseases in Man* (ed. B. Ramot), Academic Press.) Thirdly, there is the study of allozymes (the next Section and Radio programme 9, Isozymes in Man).

Protein clocks

Organisms whose proteins possess many amino acid sequences in common are likely to be more closely related than those that differ greatly in their amino acid sequences. One of the first proteins to be sequenced was haemoglobin and the similarity between haemoglobins from various vertebrates was found to be quite remarkable. Using data from several sources Zuckerandl and Pauling, for example, have estimated a mean value of 22 differences out of 141 amino acids between the α and β chains in humans and the chains with equivalent function in pigs, cattle, horses and rabbits.

Protein chemists have found that biologically active proteins, such as haemoglobin, appear to evolve at a constant rate proportional to time and not to generations, and this enables us to use comparisons between the amino acid sequences in related individuals as a measure of evolutionary lineage. Let us continue with the haemoglobin story. Fossil evidence suggests that an ancestor common to the above animals existed some 80 million years ago. This can be interpreted to mean that human haemoglobin diverged from a common molecular ancestor at a rate of one amino acid substitution per 7 million years.

QUESTION Why 7 million years and not $3\frac{1}{2}$ million?

ANSWER It is 80 million years back from humans to the common ancestor and 80 million years forward from the common ancestor to the pig. Therefore there have been 22 amino acid changes over a net period of 160 million years. The rate is given by $(22/160) \times 10^6$, which is one change in 7 million years.

If we now compare the various types of haemoglobin, we can construct an evolutionary lineage between them by relating the differences in their amino acid sequences to the above time-scale.

QUESTION The following number of amino acid substitutions have been found between the β chain and the other normal chains of haemoglobins in humans and of myoglobin in the sperm whale. Construct an evolutionary lineage map showing where such differences arose.

Myoglobin and β chain	110 out of 146 are different
α chain and β chain	76 out of 141 are different
γ chain and β chain	37 out of 146 are different
δ chain and β chain	10 out of 146 are different

ANSWER If we accept a linear time-scale of one substitution per 7 million years then the time separating the various haemoglobin chains is as shown in Figure 34.

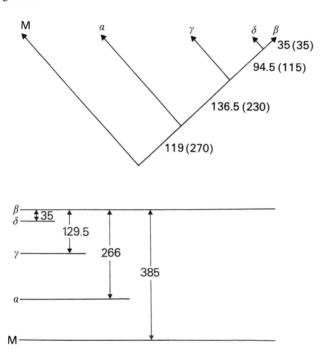

Figure 34 Evolutionary relationship between the various haemoglobins measured in millions of years. The numbers in the upper part of the Figure are obtained by subtracting adjacent numbers in the lower part of the Figure.

This calculation is not quite right because it involves some invalid assumptions. We have assumed, for example, that an amino acid can change once and only once during the time involved. Clearly, the greater the number of differences between two sequences, the greater is the chance that any particular amino acid will have changed more than once. Consequently our calculations can only be minimum estimates. Using a well-known probability theory, Pauling allowed for multiple replacements and obtained the estimates shown in parentheses in Figure 34.

But from where have they come? How is it that the α, β, γ and δ molecules, which are very similar in chemical composition and are clearly related to each other, can co-exist quite normally. We usually think of evolution as involving the replacement of one form by another; here it would appear that we have an over-all increase in the number of forms without any of the existing ones disappearing. The easiest way to begin to answer this apparent paradox is by examining, in outline, the structure of the various haemoglobin molecules which are commonly found:

HbA	$\alpha_2\beta_2$
HbF	$\alpha_2\gamma_2$
HbA$_2$	$\alpha_2\delta_2$
Hb embryonic	$\alpha_2\varepsilon_2$

HbA is the usual adult form of haemoglobin, but the other three types also all occur in normal humans. Hb embryonic is replaced in the early foetus by HbF, which is itself replaced by HbA in post-natal life. HbA$_2$ is produced after birth but at very low concentrations.

It is here that the haemoglobin variants like HbS and HbC have been so useful for the genetic analysis. From family and population studies it has been established that at least five separate loci control the synthesis of the chains and that four of these loci are in the same chromosome, with the β and δ loci closely linked.

QUESTION The genes responsible for the β and δ chains are closely linked. Of all the chains, the β and δ are the most similar. Suggest an origin for the loci concerned with haemoglobin structure.

ANSWER The simple answer to this is that each new haemoglobin chain may have resulted from a duplication of the normal haemoglobin gene, which then evolved differentially as a result of independent mutation.

Where attempts have been made to establish evolutionary relationships solely by the use of amino acid sequencing and protein chemistry, many inconsistent results have been obtained. For éxample, the β haemoglobin chain in humans bears more resemblance to that of a pig than to the β chain of the more closely related lemur. Clearly we must view amino acid sequencing as no more than a tool in the never-ending quest for 'evolutionary proof' rather than the sole criterion on which to base definitive conclusions.

10.6.3 Allozymes and genetic load   [B]

Much of the work of detecting and analysing the different haemoglobins has involved chromatography and electrophoresis. The use of the latter technique relies primarily on the minor changes in the electrical charge of a protein when one amino acid is substituted for another.

The technique in its basic form involves making an extract of the tissue or organism being studied and applying it to one point (the origin) towards the end of a specially prepared gel. The gels are most commonly made of starch, polyacrylamide or agarose and the porosity of the gel can be varied by altering the concentration of these substances. When an electric current is passed between the ends of the gel, the proteins migrate in directions related to their charge and at speeds dependent on their size, shape and charge. In this way the various proteins in the tissue are separated. Dyes are used to detect specific enzymes, and it is obvious that the same enzyme in one individual often exists in two forms. These forms are designated slow or fast according to how far they migrate in the gel. Breeding work will show whether these forms are produced by alleles or not.

Using electrophoretic techniques, Hubby and Lewontin have been able to estimate that about 30 per cent of all loci in a single population of *D. pseudoobscura* produced two or more allozymes and, by extrapolation, that individuals in the population were heterozygous for about 12 per cent of their loci. This high level of polymorphic loci had not been anticipated because it appeared contrary to concurrently held views on fitness. We need to explore the *basis* of these views before we go any further.

In 1950 Muller introduced the concept of *genetic load*. He was particularly interested in the long-term effect of radiation-induced disadvantageous mutations in humans, *and* a quantitative measurement of their effect. In his view this could be estimated by calculating the extent to which the average fitness of the population is below the fitness of the most fit genotype. Genetic load is measured in terms of reduced fitness as

genetic load

$$\frac{w_{max} - \bar{w}}{w_{max}} \tag{19}$$

$$\frac{w_{max} - \bar{w}}{w\ max}$$

where w_{max} is the fitness of the fittest genotype and $\bar{w}$ is the average fitness of the population.

In the models we developed earlier in these Units, we have taken w_{max} as unity (p. 410). Therefore, the genetic load reduces to $1 - \bar{w}$. Using the generalized model for selection derived in Section 9.6.1, we can obtain $\bar{w}$ for a diallelic polymorphism maintained by heterozygous advantage, $\bar{w} = 1 - S_1 p^2 - S_3 q^2$, and the load (often called the *segregation load* when the fittest genotype is heterozygous) is $1 - (1 - S_1 p^2 - S_3 q^2)$. Therefore the segregation load is

segregation load

$$S_1 p^2 + S_3 q^2 \tag{20}$$

$$S_1 p^2 + S_3 q^2$$

Check that the relationship '$\bar{w} = w_{max}$ − segregation load' is correct for heterozygous advantage.

Now let us see why the results of Hubby and Lewontin were a bombshell to the concept of genetic load. They estimated that some 30 per cent, or 2 000, of the loci in

484

D. pseudoobscura were heterozygous. If we assume that the alleles occur at equal frequencies and that there is 1 per cent selection against each of the homozygotes (that is, $S_1 = S_3 = 0.01$), the segregation load per locus is

$$(0.01 \times 0.5^2) + (0.01 \times 0.5^2) = 0.005$$

and

$$\bar{w} = 1 - 0.005 = 0.995$$

Check that if we take any other legitimate values of p and q other than 0.5 (for example 0.9 and 0.1), then the segregation load will be *larger* than 0.005.

If we now assume that the average fitness of *each* of the 2 000 polymorphic loci is 0.995 and that the other 70 per cent (4 660) of the loci have *unit* fitness, then the over-all fitness of the whole population would be

$$(0.995)^{2\,000} \times 1^{4\,660} = 0.000\,044 \times 1 = 0.000\,044$$

An average fitness as low as 0.000 044 effectively means that the population does not exist! Clearly this is a nonsense because Hubby and Lewontin were studying a viable and essentially normal population of *D. pseudoobscura*, and hence we must conclude that a genetic load as small as 0.005 is too large in the example we have taken!

QUESTION Suggest two possibly erroneous assumptions that we have made.

ANSWER (a) Perhaps it is wrong to consider the polymorphic loci separately and hence the concept of genetic load is not relevant in this context.
(b) The selective coefficients of 0.01 are possibly too high for most loci.

When we calculated the mean fitness of the population, we accumulated the effects of selection acting on each of 2 000 loci as if they were independent events. It is more reasonable to assume that alleles at different loci interact in favourable or unfavourable ways to produce the phenotype, and, consequently, alleles are more likely to be subject to selection as members of a co-adapted gene complex rather than as isolated units. Under this condition, the genetic load could be far lower than we earlier envisaged*. (See also the history text†, Section H.6—classical and balance theories.)

10.6.4 Allozymes and the neutral-allele hypothesis

The above conclusion is no doubt valid, but in many instances the second answer to the last question may be important, and this is the one that we shall now consider. If we revert to our calculation of the over-all fitness of the population (that is, to $(0.995)^{2\,000} \times 1^{4\,660}$), we have only to assume that the alleles at 1 000 of the polymorphic loci are selectively neutral with respect to each other for the over-all fitness to become

$$(0.995)^{1\,000} \times 1^{5\,660} = 0.006\,6$$

which, although small, is much more reasonable.

Alternatively we can recalculate the fitness, assuming a smaller average selective coefficient per locus. Thus, if $S_1 = S_3 = 0.001$

$$\bar{w} = 1 - (0.001 \times 0.5^2) + (0.001 \times 0.5^2)$$
$$= 1 - (0.002 \times 0.25)$$
$$= 1 - 0.000\,5$$
$$= 0.999\,5$$

and the over-all fitness is $(0.999\,5)^{2\,000} \times 1^{4\,660} = 0.368$, which is even more respectable.

* Do not, however, now conclude that selection never acts differentially on alleles at a single locus. There is ample evidence that it does on occasion, and several examples have already been given in this Course.

† The Open University (1976) S299 HIST *The History and Social Relations of Genetics*, The Open University Press. This text is to be studied in parallel with the Units of the Course.

We could go on reducing the genetic load still further by assuming even smaller selective coefficients, but already, with selective coefficients at 0.001, the mutation rate is becoming important. We saw in Section 10.3 that the balance between a selective coefficient of 0.001 and a mutation rate of 10^{-5} per generation can maintain a disadvantageous recessive allele in the population at a frequency of 10 per cent! Furthermore, in the absence of selection, genetic drift becomes an important force for changing allele frequencies in small populations (see Section 10.5.3), leading quite arbitrarily to the fixation of some alleles and the consequential loss of others.

It is by using rather more sophisticated arguments along these lines that Kimura and his colleagues suggested that allozymes were more often than not determined by neutral alleles and would, therefore, represent transient polymorphic states produced by neutral alleles and genetic drift.

The theory of neutral alleles suggests that at any one time, many genes will be in the process of allelic substitution, and will, therefore, be polymorphic. In the absence of selection, the distribution of the different morphs should not correlate with any climatic, edaphic* or geographic variables. It is, however, extremely difficult to detect neutral alleles, since we cannot always evaluate small differences in selection pressures that may operator in favour of, or against alleles at the same locus. The inability to detect such selection pressures does not mean, however, that they do not occur. Remember the question and answer at the end of Section 10.6.1.

Another problem with the interpretation mainly results from the techniques used. Gel electrophoresis is a very sensitive technique for separating very small quantities of closely related enzymes, and, although some enzymes are routinely assayed in individuals of *Drosophila*, man and mouse, these enzymes often occur at concentrations below the level at which their activity can be estimated reliably. Consequently, it is much easier to show that the biochemical variation exists, than it is to estimate its significance. Moreover, and this is a particularly important point, the enzymes are detected on the gel by specific histochemical tests that demonstrate the *qualitative activity* of the enzyme. They can tell you nothing about how the enzymes act in the living organism. It may be that in most normal conditions the allozymes show no important differential activities that have any effect on fitness. Under stress, however, or when the environmental conditions fluctuate or oscillate during the life of the organism, it may well be a different matter. It has been argued, for example, that the intensity of selection will change with the density of individuals in a population. That is, there is an ecological load, which gives rise to an *apparent* genetic load. This genetic load mainly results from the situation in which not all genotypes are equally viable when the population becomes crowded. Many individuals have to die in the process of density regulation, and if those that die differ genetically from those that survive we shall observe a genetic load. This load will come and go according to the density of the individuals, and can still be much higher than we would expect, without any danger that the population would become extinct. We shall not consider density-dependent selection any further in this Course.

Furthermore, if an enzyme normally occurs as a dimer, then a heterozygous individual may contain *three* and not two forms of the enzyme. The third type is a hybrid molecule. Fincham and his colleagues have described such an enzyme, glutamate dehydrogenase, in *Neurospora* and they suggest that the ability of the heterokaryon to produce three types of this enzyme may enable it to be better suited than homokaryons to fluctuating environments.

The problem for the person who believes that selection is of prime importance is to demonstrate that certain stress conditions *do* have a differential effect on the allozyme and that such conditions are of importance in natural habitats.

Kimura, on the other hand, asserts that the models he has developed already demonstrate that most of the allozyme variation is neutral and hence that he does not have to demonstrate this in actuality.

In 1968 Kimura developed a model that showed that the rate of fixation by drift was equal to the average mutation rate; that is, the rate of fixation, measured in terms of the numbers of neutral alleles being fixed per generation, was equal to μ, the rate per generation at which the neutral alleles were appearing at each locus by mutation.

We shall now see how he reached this conclusion.

* Edaphic: pertaining to variation in the structure and composition of the soil.

Let the population size be equal to N. Then the number of genes at a particular locus in the population $= 2N$. Let the mutation rate be equal to μ.

Kimura argued that once a neutral allele comes into being, it occupies $1/2N$ of all the genes at that particular locus within the population. Since it is neutral, it has the same chance as every other allele present at the time of having its descendants fixed in all $2N$ positions. In other words, the chances that the descendants of a particular neutral mutant will be fixed to the exclusion of all other alleles is $1/2N$ (Section 10.5.2). It follows that the probability of a neutral allele being fixed is the product of the number of new alleles arising in any one generation ($2N\mu$), and the probability of one allele occupying all $2N$ positions ($1/2N$); that is

$$2N\mu\left(\frac{1}{2N}\right)$$

which reduces to μ. Therefore the fixation rate is equal to the mutation rate.

In Section 10.6.1 we discussed the evolution of the various forms of haemoglobin. Evidence suggested that the rate of amino acid substitution was 1 per 7 million years. Assuming that there are about 146 separate amino acids per chain, the substitution rate per chain is calculated as one per $7 \times 10^6 \times 146$, which equals one per $1\,022\,000\,000$ years, that is approximately 10^{-9} substitutions per chain per year.

This amount of evolution, which is significant when we consider all the genes for haemoglobin, can be accounted for entirely by the fixation of neutral alleles. This, of course, does not mean to say that the evolution of haemoglobin has occurred or is occurring entirely by the fixation of neutral alleles. If we examine the mutation rate of haemoglobin, whose value is of the order of 10^{-5}, it is clear that evolution is more conservative than the neutral mutation theory suggests, and hence some new mutations are at a selective disadvantage. When one considers the gross abnormalities and selective disadvantage of the single amino acid substitution in sickle-cell anaemia homozygotes, it is not surprising that the rate of allele substitution is far less than the total mutation rate.

Up to now in our discussions of allozyme polymorphism and genetic load we have not considered frequency-dependent selection. Let us examine what effect this will have on the genetic load. We shall consider one example very briefly and then state some general conclusions drawn from more detailed studies.

On p. 462 we obtained equation 15 for the mean fitness of a frequency-dependent situation without dominance

$$\bar{w} = 1 - S(p^4 + 4p^2q^2 + q^4)$$

If we take $S = 0.01$ and $p = q = 0.5$ (which are exactly the same values as those that we substituted into the equation for $\bar{w}$ on p. 485), we obtain

$$
\begin{aligned}
\bar{w} &= 1 - 0.01(0.062\,5 + 4 \times 0.062\,5 + 0.062\,5) \\
&= 1 - (0.01 \times 0.062\,5 \times 6) \\
&= 1 - 0.003\,75 \\
&= 0.996\,25
\end{aligned}
$$

Now what is w_{max} in this situation? We saw on p. 463 that at equilibrium the fitness of the two homozygotes was $1 - S/4$ and the fitness of the heterozygote was $1 - S/2$. Therefore, w_{max} in this case is $1 - S/4$. Substituting these values of $\bar{w}$ and w_{max} in expression 19, we obtain

$$\frac{1 - (0.01/4) - 0.996\,25}{1 - 0.01/4} = \frac{0.001\,25}{0.997\,5} = 0.001\,253\,13$$

This gives us a genetic load of $0.001\,253\,13$, but we cannot compare this load *directly* with the value (0.005) we obtained on p. 485 for heterozygous advantage. This is because the value $0.001\,253\,13$ relates to a maximum fitness of $1 - S/4$, whereas the load of 0.005 relates to a maximum fitness of unity. We can change the scale by dividing the load by the appropriate w_{max}. Thus, $0.001\,253\,13/0.997\,5 = 0.001\,256\,27$. This is *smaller* than the segregation load of 0.005, and corresponds with a mean fitness per locus of $0.998\,743\,73$. The average fitness for the *Drosophila* example is now given by $0.998\,743\,73^{2\,000} \times 1^{4\,660} = 0.080\,93$. This mean fitness is nearly $2\,000$ times larger than that for the model of heterozygous advantage.

You may well have been puzzled why we have been working to eight places of decimals! It is easily seen that it is the sixth place which is the crucial one and we need two more at least to avoid rounding-off errors. This is very important here because we have a number raised to the power 2 000, so any errors are magnified enormously.

Other studies have shown that under some conditions, not only is the genetic load reduced but it may disappear altogether. Thus loci that are subjected to frequency-dependent selection can be polymorphic without making any contribution to the genetic load. Frequency-dependent selection could therefore be alternative to or additional to the neutral-allele hypothesis. Clearly this is a complicated topic, but it is one that is fundamental to our understanding of population genetics. It is not surprising, therefore, that some very able people in different parts of the world are discussing, arguing and even quarrelling over their attempts to understand what is going on.

Summary

1   The discovery that many organisms are much more variable than was originally supposed has led to a re-examination of some of the basic concepts of population genetics.

2   One explanation is that much of the allozyme polymorphism observed in natural populations results from the genetic drift of neutral alleles.

3   On the other hand, other people are not satisfied that any allele can be neutral for more than a few generations, if at all, and consequently, that selection, density- and frequency-dependent effects are the major cause of the allozyme polymorphisms.

You should now attempt SAQ 11.

Summary of Units 9 and 10

By rereading the summaries at the ends of each Section you will be able to make your own summary of these Units, but we shall add a postcript.

The genetics of populations involves the study of variation in natural habitats together with laboratory experiments and theoretical model building. In the early days of the subject, the arguments of Fisher, Haldane and Wright gave great insight into the basic problems and clearly defined the terms of reference within which the answers should be found. Some of the original premises have subsequently been found to be too simplistic, essentially because factors that were originally thought to be of little or no consequence have turned out, on closer inspection, to be very important under certain circumstances. Furthermore, entirely new phenomena have been discovered as a result of work on natural populations of animals and plants.

We have a theoretical framework that is essentially mathematical and we have much data collected from experimentation and observation, both in the field and in the laboratory. Currently we are trying to synthesize an over-all view of ecological and population genetics.

This is as far as we can take our study of discontinuous variation; we shall go on to examine continuous variation in Units 11–13.

Self-assessment questions

SAQ 1 Derive the Hardy–Weinberg equilibrium for the frequency of a pair of alleles with three genotypes and list the known assumptions implicit in the model.

SAQ 2 Give three reasons why allele frequencies are unlikely to attain stable equilibria by mutation pressures of 1×10^{-5}.

SAQ 3 During 1961 Parsons and colleagues collected male *D. melanogaster* at Newnham, Cambridge and screened the samples for the frequency of the ebony allele e^{New}. On 8 July, they obtained the following numbers of the three genotypes.

$e^+ e^+$	$e^+ e^{New}$	$e^{New} e^{New}$
15	56	9

Assuming that these frequencies were the same as those in the adults of the previous generation:

(a) Are these genotype frequencies in Hardy–Weinberg equilibrium?

(b) Estimate the relative fitness of the three genotypes.

(c) Interpret the results of your analysis and explain its limitations.

SAQ 4 What were the two crucial discoveries made by Kettlewell at an early stage of his work on industrial melanism in *B. betularia*?

SAQ 5 Following an experiment he began with Holt on the dominant mutation called Danforth's short-tail (*Sd*) (see Fig. 35) in the house mouse (*Mus musculus*), Fisher designed an experiment to select for increase and decrease in tail length.

(a)

(b)

Figure 35 (a) Normal mouse; (b) mouse showing Danforth's short-tail mutation.

It had been shown previously that (1) *Sd Sd*⁺ mice had kinky and short tails caused by skeletal defects, (2) no mice of the genotype *Sd Sd* reached maturity and (3) approximately one quarter of the progeny of *Sd Sd*⁺ × *Sd Sd*⁺ matings had no tails and died shortly after birth of urinogenital disorders.

Fisher's experiment involved mating heterozygous individuals, choosing in one line individuals with the longest tails as the parents of the next generation and in the other line individuals with the shortest tails. After some initial success, the line in which Fisher was selecting for short tails died out. On autopsy, these mice were found to have urinogenital disorders. In the line in which he was selecting for long tails, he found after six generations that (1) he had difficulty in distinguishing between mice with *Sd* and those with wild-type tails, and (2) some of the mice born without tails did not die. Indeed, several females were weaned successfully and reached maturity. Moreover, he found that when he mated these tail-less mice with long-tailed (wild-type) males *all* the progeny had short kinky tails.

Are the following statements true or false?

(a) When Fisher was selecting for long tails he was selecting for dominant expression of the *Sd* allele.

(b) By selecting for short tails, Fisher was unintentionally, but necessarily, also selecting for accentuation of the urinogenital disorders.

(c) *Sd Sd* homozygotes were capable of reproduction at the beginning of the experiment.

(d) *Sd Sd* homozygotes were capable of reproduction by the end of the experiment.

(e) The experiment demonstrates that by starting with one allele showing incomplete dominance it is possible to select for both dominant and recessive expression of that allele with reasonable success.

SAQ 6 Select the correct figure, letter or term in the following statements.

(a) Hubby and Lewontin estimated that the average population of *D. pseudoobscura* is polymorphic for about 40%   35%   30%   25%   20%   15% of all loci.

(b) Horowitz collected 12 strains of *N. crassa* from the wild and found 1   2   3   4   5   6 different forms of tyrosinase.

(c) The work of Jürgensen and Schwarz suggested that people of blood group A_1 A_2 A_1B A_2B B   O   were most likely to live to a healthy old age.

(d) Vogel and his colleagues used   random samples   stratified samples sibs   systematic samples   for their controls when working on the association between smallpox and the ABO blood groups in India.

SAQ 7 (a) Is it possible to determine the genotype of an individual *C. hortensis* (see notes for TV programme 9) merely by looking at its phenotype?

(b) Assume that *a, b, c* represent the frequencies of brown, pink and yellow shells in a sample of *C. nemoralis* and that the frequency of the brown, pink and yellow alleles can be estimated from

$$\begin{aligned} \text{yellow} \quad & r = \sqrt{c} \\ \text{pink} \quad & q = \sqrt{(b + c)} - \sqrt{c} \\ \text{brown} \quad & p = 1 - \sqrt{(b + c)} \end{aligned}$$

We have collected a sample containing 39 browns, 45 pinks and 16 yellows. What are the allele frequencies?

SAQ 8 (a) What were the three main conclusions to be drawn from van Zandt Brower's experiment with the monarch, *D. plexippus*, and the viceroy, *L. archippus archippus*?

(b) For Batesian mimicry to work, what necessary characteristics must a predator possess?

(c) In what way does the experiment of Birch, using an excess of food, help to explain the annual fluctuations in the frequency of the CH arrangement of chromosome III in *D. pseudoobscura* at Piñon Flats?

SAQ 9 Which of the following statements are true of gametophytic and/or sporophytic incompatibility systems?

Assign the appropriate letter of the key to each statement.

Key:

A   Gametophytic systems only.

B   Sporophytic systems only.

C   Both gametophytic and sporophytic systems.

D   Neither gametophytic nor sporophytic systems.

Statement:

1   The pollen reaction is directly determined by the pollen parents' diploid genotype.

2   The pollen reaction is determined by the paternally originating cytoplasm of the pollen grain.

3   Semi-compatible crosses are possible.

4   The system is determined by a physiological reaction.

5   Homozygotes may be produced as the result of a cross.

SAQ 10 A tropical insect was found to show sexual dimorphism as regards life history. For example, in one isolated population females required an average of 4 months to reach reproductive age, whereas the males matured more rapidly and reproduced at a mean age of 2 months. Generations were found to be continuous and overlapping with an estimated 50 males and 100 females achieving sexual maturity every month. If a new mutation arises in this population, what is the criterion for establishing whether its future will be determined primarily by selection or whether it will be effectively neutral with a future determined primarily by random genetic drift?

SAQ 11 Fibrocystic disease of the pancreas (FCD or cystic fibrosis) is the most common genetically transmitted disease in northern Eurasia and North America. The inheritance of the disease shows the typical autosomal recessive pattern and about 1 in 2 000 births are of affected children.

(a) Estimate the frequency of those individuals who are heterozygous 'carriers' of the recessive allele (A_2) for cystic fibrosis.

(b) Estimate the rate of mutation from the normal allele (A_1) required to maintain FCD at its observed frequency.

Assume that the disease is effectively lethal. Comment on the likelihood of such a mutation rate being found in nature.

(c) If carriers enjoy a selective advantage, estimate the relative fitnesses of the three genotypes necessary to maintain FCD at its observed frequency. Again assume that the disease is effectively lethal.

(d) By comparing the fecundities of $A_1A_1 \times A_1A_2$ and $A_1A_1 \times A_1A_1$ marriages it has been estimated that the fitness of A_1A_1 individuals may be as low as 0.9 compared with the heterozygotes. If this is so, estimate the potential frequency of affected births and the segregational load should the frequency of the A_2 allele come to an equilibrium as a result of this heterozygous advantage. Interpret your answers.

Answers to ITQs

ITQ 1 (*Objective 4 and part of Objective 2*) The procedure follows essentially the same pattern as the one used for Table 6. Thus we have the mating and progeny frequencies indicated in Table 22. The genotype frequencies among the progeny are those stated in the question.

Table 22 Mating and progeny frequencies for ITQ 1

Parents		Frequency of mating	Frequency of progeny		
♀	♂		A_1A_1	A_1A_2	A_2A_2
A_1A_1	A_1A_1	0.68×0.68	0.462 4		
A_1A_1	A_1A_2	0.68×0.24	0.081 6	0.081 6	
A_1A_1	A_2A_2	0.68×0.08		0.054 4	
A_1A_2	A_1A_1	0.24×0.68	0.081 6	0.081 6	
A_1A_2	A_1A_2	0.24×0.24	0.014 4	0.028 8	0.014 4
A_1A_2	A_2A_2	0.24×0.08		0.009 6	0.009 6
A_2A_2	A_1A_1	0.08×0.68		0.054 4	
A_2A_2	A_1A_2	0.08×0.24		0.009 6	0.009 6
A_2A_2	A_2A_2	0.08×0.08			0.006 4
Totals		1.000 0	0.640 0	0.320 0	0.040 0

Note that this result demonstrates that even if one starts with genotype frequencies that are not at Hardy–Weinberg equilibrium values, the equilibrium frequencies are achieved after *one* cycle of mating.

ITQ 2 (*Objective 4 and part of Objective 5*) On counting the alleles we find that there are 450 spotting alleles (Sp) and 550 non-spotting alleles (sp), and so $p = 0.45$ and $q = 0.55$. The expected frequency of the $Sp\,Sp$ genotype is $p^2N = 0.45 \times 0.45 \times 500 = 101.25$. Similarly, the expected frequencies of the other genotypes can be calculated and arranged in a table (see Table 23).

Table 23

Genotype	Frequency		$O - E$	$(O - E)^2/E$
	Observed	Expected		
$Sp\,Sp$	100	101.25	-1.25	0.015 43
$Sp\,sp$	250	247.50	$+2.50$	0.025 25
$sp\,sp$	150	151.25	-1.25	0.010 33
Totals	500	500.00	0.00	0.051 01

We have obtained a value of $\chi^2 = 0.05$, but how many degrees of freedom are attached to this χ^2? At first sight you might say two, but we have used the data to obtain the allele frequencies and then used these to calculate the expected genotype frequencies. We have used one degree of freedom to do this and so our χ^2 test of significance has only one degree of freedom attached. Thus $\chi^2_{[1]} = 0.05$, and so $P > 0.80$.

We can conclude that there is no evidence to suppose that the observed genotype frequencies are not in Hardy–Weinberg equilibrium.

ITQ 3 (*Objective 2*) Allele frequencies remain constant from generation to generation; that is, an equilibrium in allele frequencies is maintained. If you had difficulty with this ITQ, return to Section 9.4.2.

ITQ 4 (*Objective 4*) Using exactly the same procedure as outlined in the answer to ITQ 2, we obtain $p = 0.43$ and $q = 0.57$ and hence the results shown in Table 24.

Table 24

Genotype	Frequency		$O - E$	$(O - E)^2/E$
	Observed	Expected		
$Sp\,Sp$	80	92.45	-12.45	1.676 61
$Sp\,sp$	270	245.10	$+24.90$	2.529 62
$sp\,sp$	150	162.45	-12.45	0.954 16
Totals	500	500.00	00.00	5.160 39

We have obtained $\chi^2_{[1]} = 5.16$ and thus $0.05 > P > 0.01$. We must conclude that the observed frequencies are not in Hardy–Weinberg equilibrium.

ITQ 5 (*Objective 12*) In special case (d) (p. 423) it was shown that it was possible to maintain a pair of alleles in a population when the heterozygote is the fittest genotype. Furthermore, when the relative fitnesses of the two homozygotes are more or less constant, but not necessarily the same, a stable equilibrium is possible.

ITQ 6 (*Part of Objective 10*) The three homozygotes $C^B C^B$, $C^P C^P$ and $C^Y C^Y$ will have genotype frequencies p^2, q^2 and r^2. The three heterozygotes $C^B C^P$, $C^B C^Y$ and $C^P C^Y$ will have genotype frequencies $2pq$, $2pr$ and $2qr$, respectively. The expected phenotype frequencies are then as shown below.

Phenotype	Frequency
brown	$p^2 + 2pq + 2pr$
pink	$q^2 + 2qr$
yellow	r^2

ITQ 7 (*Part of Objective 4 and part of Objective 10*) The phenotypic frequencies in the sample are 0.19 brown ($=a$), 0.45 pink ($=b$) and 0.36 yellow ($=c$). The allele frequencies are, therefore, calculated as

$$\text{brown} \quad p = 1 - \sqrt{(b + c)} = 0.1$$
$$\text{pink} \quad q = \sqrt{(b + c)} - \sqrt{c} = 0.3$$
$$\text{yellow} \quad r = \sqrt{c} = 0.6$$

ITQ 8 *Objective 9 and Objective 12*) On crossing allopatric races the mimicry breaks down. On crossing sympatric races the mimicry is maintained because of complete dominance between the female forms (see also Section 9.7.2).

ITQ 9 (*Part of Objective 11*) If the male is heterozygous there is a chance of $\frac{1}{2}$ that a particular chromosome will appear in a particular larva. Having scored that chromosome, we must ask how many more larvae must we test to have a chance of only $1/64$ that they are all the same as the first one. The probability is $\frac{1}{2}$ that it will be the same each time we score a larva. Thus in n *subsequent* trials the chance that all the chromosomes will be the same as the first is $(\frac{1}{2})^n$. But we have already said that the over-all chance must not be greater than $\frac{1}{64}$. Therefore we write $(\frac{1}{2})^n = \frac{1}{64}$ and solve the equation. Taking logs

$$n \log \tfrac{1}{2} = \log \tfrac{1}{64}$$

$$-n \log 2 = -\log 64$$

$$n = \frac{\log 64}{\log 2} = 6$$

Therefore the total number of larvae that need be examined is 7 (6 + the first one).

Although you may have seen at once that $64 = 2^6$, the method of calculating n by taking logs has been used here because it is often the only way of solving similar problems which occur not infrequently in the design of genetic experiments. A good example of the method will be met in Section 10.4.2, where we need to solve $(1 - m)^{10} = 0.82$.

ITQ 10 (*Part of Objective 2*) (a) The mating of related individuals.

(b) An increase in the frequency of homozygotes.

If you had difficulty with this question, refer back to Section 9.1.1 for (b) and to Section 9.4.4 for (a).

ITQ 11 (*Course Objective 3(a)*) These genes must be located in segment II.

ITQ 12 (*Objective 13*)

(i) (a) Not applicable.

(b) $S_1S_2 \times S_2S_3 \rightarrow \frac{1}{2}S_1S_3 : \frac{1}{2}S_2S_3$.

(c) $S_1S_2 \times S_3S_4 \rightarrow \frac{1}{4}S_1S_3 : \frac{1}{4}S_1S_4 : \frac{1}{4}S_2S_3 : \frac{1}{4}S_2S_4$.

(ii) (a) Not applicable.

(b) Both types of progeny are semi-compatible with the female parent. One type of progeny (S_1S_3) is semi-compatible with the male parent, the other is fully incompatible.

(c) All progeny are semi-compatible with both parents.

(iii) (a) Not applicable.

(b)

		Male parents			
		S_1S_3		S_2S_3	
	Pollen	S_1 S_3		S_2 S_3	
					Pollen reaction
Female parents	S_1 S_3	$-$ $-$		$+$ $-$	$+ =$ compatible
	S_2 S_3	$+$ $-$		$-$ $-$	$- =$ incompatible

Half of the sib-matings are fully incompatible and half are semi-compatible.

| | | Male parents | | | | | | | |
| | | S_1S_3 | | S_1S_4 | | S_2S_3 | | S_2S_4 | |
	Pollen	S_1	S_3	S_1	S_4	S_2	S_3	S_2	S_4
Female parents $\quad S_1 \; S_3$		−	−	−	+	+	−	+	+
$S_1 \; S_4$		−	+	−	−	+	+	+	−
$S_2 \; S_3$		+	−	+	+	−	−	−	+
$S_2 \; S_4$		+	+	+	−	−	+	−	−

$\frac{1}{4}$ fully incompatible : $\frac{1}{2}$ semi-compatible : $\frac{1}{4}$ fully compatible

(iv) A trick question. There is no way, barring mutation, that a plant with this kind of incompatibility system can be homozygous for an S allele.

(v) The minimum number is three. If there were only two, all plants would be S_1S_2 and all matings would be fully incompatible.

ITQ 13 (*Objective 13*)

	Style Pollen	Style Pollen
Genotype	$S_1S_3 \quad S_3S_4$	$S_3S_4 \quad S_1S_3$
	×	×
Phenotype	$S_1S_3 \mid S_3$	$S_3 \mid S_1$
	↓	↓
	Fully incompatible	Fully compatible
Progeny	None	$\frac{1}{4}S_1S_3 : \frac{1}{4}S_1S_4 : \frac{1}{4}S_3S_3 : \frac{1}{4}S_3S_4$

ITQ 14 (*Objective 3 of Unit 1*) The fact that pin intercrosses breed true, whereas the progeny of thrum intercrosses segregate to give three thrums to one pin, suggests that the character is determined by a pair of alleles at a single locus. Pins must be homozygous for the recessive allele and thrums heterozygous. This interpretation is confirmed by the 1 : 1 segregations from the reciprocal crosses between pin and thrum. The gene symbols S and s are usually used for the two alleles, so that the genotypes of thrums and pins are respectively Ss and ss.

ITQ 15 (*Objective 13*) The form of the thrum flower is such that thrum pollen must often fall down onto thrum stigmas. If thrums were self-compatible, more self-fertilization would undoubtedly occur. Pin flowers, because the stigma is above the anthers, do not suffer from this problem.

ITQ 16 (*Course Objective 3(a)*) The EGG. The principal reason why a geneticist will argue that the egg came first is that the egg is the only stage in the life cycle at which the chicken exists as a single cell. The chicken develops by the division of the fertilized egg. You may be thinking: but the chicken has to lay the egg first. True, but the first chicken came from a fertilized egg containing mutation/s that neither its mother nor its father possessed. In other words neither of the parents of the first chicken was a chicken!

ITQ 17 The Hardy–Weinberg law. If you could not answer this question, you should reread Section 9.4.2.

ITQ 18 (*Objectives 2 and 4*) This is essentially the same kind of problem as the one we solved in the last Question/Answer in Section 9.4.2. All we have to do is substitute $q = 0.707$ in the standard binomial expansion of $(p + q)^2$

$$\begin{array}{ccc} A_1A_1 & A_1A_2 & A_2A_2 \\ p^2 & 2pq & q^2 \end{array}$$

We cannot distinguish A_1A_1 from A_1A_2, and hence the frequency of the dominant and recessive phenotypes at equilibrium are 0.5 : 0.5, respectively.

ITQ 19 (*Objective 16*) Doubling the mutation rate means that the value of q^2 at each step in the Table on p. 465 must be doubled. Hence we obtain the results given in the following Table.

S	0.001	0.01	0.1	0.5	1.0
q^2	0.02	0.002	0.000 2	0.000 04	0.000 02
q	0.141 4	0.044 72	0.014 14	0.006 325	0.004 472

ITQ 20 (*Objectives 4 and 16*) In the Question and Answer immediately preceding this ITQ we found that $q = 0.001$.

If there is only one individual who is a recessive homozygote then

$$q^2 = \frac{1}{N}$$

where N is the population size (see Section 9.4.2 if you have forgotten this), and

$$N = \frac{1}{q^2}$$

$$\therefore N = \frac{1}{10^{-6}} = 10^6$$

Thus for one recessive homozygous individual to be produced each generation, the population must contain one million individuals!

ITQ 21 (*Objective 18*) The mean frequency of the PP chromosome arrangement shows a greater variance among the populations derived from the small initial founder populations.

Answers to SAQs

SAQ 1 (*Objective 2*) Reread Sections 9.4.2 and 9.4.4.

SAQ 2 (*Objective 3*)

(a) A stable equilibrium is likely to be attained by this mutation pressure only when the population is larger than 25 000 breeding individuals.

(b) It would take too long to change allele frequencies towards their equilibrium values.

(c) If mutation were the only process in operation, populations would become full of strange and bizarre mutant forms. It is patently obvious that this is not so, and hence we must conclude that mutation pressure alone is not sufficient to account for evolutionary change.

Reread Section 9.5.3 if you had difficulties.

SAQ 3 (*Objectives 4, 5, 6 and 7*)

(a) The genotype frequencies are

	e^+e^+	e^+e^{New}	$e^{New}e^{New}$	Total
	15	56	9	80

We calculate the frequency of the e^+ allele as

$$\frac{(2 \times 15) + 56}{160} = 0.537\,5$$

Assuming random mating and that the progeny are initially produced in binomial proportions $p^2 + 2pq + q^2$, we proceed to calculate the expected genotype frequencies, as follows.

	Observed	Expected
e^+e^+	15	23.11
e^+e^{New}	56	39.78
$e^{New}e^{New}$	9	17.11
Totals	80	80.00

$$\chi^2_{[1]} = \frac{(15 - 23.11)^2}{23.11} + \frac{(56 - 39.78)^2}{39.78} + \frac{(9 - 17.11)^2}{17.11}$$

$$\chi^2_{[1]} = 13.31$$

Therefore the probability that this sample is in Hardy–Weinberg equilibrium is less than 0.001. We can conclude, therefore, that it is significantly disturbed from equilibrium.

(b) We now proceed to estimate the relative fitness of the genotypes. It is clear that the heterozygote is greatly in excess of expectation and so let us ascribe to it a relative fitness of 1. We can write, therefore:

	e^+e^+	e^+e^{New}	$e^{New}e^{New}$	Total
Before selection	23.11	39.78	17.11	80
After selection	15	56	9	80
Relative fitness	$1 - S_1$	1	$1 - S_3$	

The relative fitness of e^+e^+ is therefore

$$\frac{15 \times 39.78}{23.11 \times 56} = 0.461$$

and the relative fitness of $e^{New}e^{New}$ is

$$\frac{9 \times 39.78}{17.11 \times 56} = 0.374$$

Thus in our model the relative fitnesses are expressed as follows.

e^+e^+	e^+e^{New}	$e^{New}e^{New}$
$1 - S_1$	1	$1 - S_3$
$1 - 0.539$	1	$1 - 0.626$
0.461	1	0.374

We can check our calculation by working the other way round; that is:

	e^+e^+	e^+e^{New}	$e^{New}e^{New}$	Total
Expected on binomial proportions	23.11	39.78	17.11	80
Relative fitness	$1 - 0.539$	1	$1 - 0.626$	
Relative frequency after selection	10.65	39.78	6.40	56.83

We can now use these relative frequencies to estimate the expected numbers of the three genotypes in a sample of 80. We multiply each frequency by 80/56.83 and obtain 14.99, 55.99 and 9, which are the same numbers as Parsons observed in his sample.

(c) 1   It appears that the polymorphism of ebony body colour in the Newnham population of *D. melanogaster* is maintained by heterozygous advantage.

2   Although the conclusion above is a very reasonable one to draw from the data given, there are problems over its meaning.

(i) We know nothing about the selective agents involved.

(ii) We know nothing about the general ecology of the population.

(iii) The way in which the question was worded implied that Parsons had sampled on several occasions during the summer of 1961. He did, and so he was able to judge that the sample he obtained on 8 July was not unusual and bore a reasonable resemblance to samples collected earlier and later. We have not emphasized the point earlier, because it is mentioned in *STATS*, but it is very important that more than one sample of any particular population being studied should be taken, because one then has an estimate of *sampling error*.

(iv) We know nothing about the origin of these populations or about migrations of flies into or out of the population.

(v) From the way the problem was worded, we can guess that ebony is not sex linked, but are the generations discrete or are they overlapping?

It is clear that we have a polymorphism that needs detailed study before any sensible comments can be made, other than that it appears to be maintained by heterozygous advantage.

SAQ 4 (*Objective 8*)

(a) The different results obtained from the experiments of mark, release and recapture of moths in Dorset and in Birmingham.

(b The proof that the less-well-camouflaged form was more likely to be eaten by birds than the better camouflaged morph.

SAQ 5 (*Objective 9*)

(a) False. He was selecting for recessive expression of the *Sd* allele.

(b) Probably truc, but more evidence is needed.

(c) False. It is likely that the progeny of $SdSd^+ \times SdSd^+$ matings that had *no* tails and died young were *SdSd* individuals.

(d) True. Tail-less mice crossed with wild type gave all kinky-tailed progeny; that is

$$SdSd \times Sd^+Sd^+$$

$$\downarrow$$

$$SdSd^+$$

(e) True. But note that the experiment did not go to completion in either direction.

SAQ 6 (*Objectives 11 and 12*)

(a) 30 per cent. See Section 10.1.

(b) 4. See Section 10.1.1.

(c) Group O. See Table 3.

(d) Sibs. See Table 4.

SAQ 7 (*Objective 10*)

(a) No. See answer to Question on p. 437.

(b) $r = \sqrt{0.16} = 0.40$;
   $q = \sqrt{(0.45 + 0.16)} - \sqrt{0.16} = 0.38$;
   $p = 1 - \sqrt{(0.45 + 0.16)} = 0.22$.

SAQ 8 (*Objectives 11, 12 and 14*)

(a) 1 Mimicry works.
2 Birds *vary* in their ability to distinguish mimics from models.
3 Mimicry does not have to evolve in a single step. Even a poor mimic has *some* protection.

(b) The predators must have colour vision and reasonable memories.

(c) See p. 446.

SAQ 9 (*Objective 13*)

1 B;   2 B;   3 A;   4 C;   5 B.

SAQ 10 (*Objective 18*) We must first determine the effective population size, allowing for overlapping generations. As the two sexes differ in life history, they must, initially, be treated independently. An effective population number may be obtained for each sex by applying the equation

$$N = tN^*$$

where t is the mean age in months at reproduction and N^* is the number of males or females who reach reproductive age each month.

We have, therefore

$$\text{males} \quad N_m = 2 \times 50 = 100$$
$$\text{females} \quad N_f = 4 \times 100 = 400$$

Finally, the effective size of the whole population may be obtained from

$$N_e = \frac{4N_m N_f}{N_m + N_f} = \frac{4 \times 100 \times 400}{500} = 320$$

In order to determine whether or not a new mutation will be effectively neutral it is necessary to estimate $4N_e S$, where S is the selective coefficient associated with the new allele. If $4N_e S < 1$, the allele is effectively neutral and its future will be determined primarily by random genetic drift. In this case, for $4N_e S < 1$ or $S < 1/4N_e$, we must have S less than 0.000 8.

SAQ 11 (*Objectives 4, 6, 15 and 20*)

(a) Let q = the frequency of A_2, the FCD allele and

   p = the frequency of A_1, the normal allele.

498

Assuming that the Hardy–Weinberg law may be applied

$$q^2 = \frac{1}{2\,000}$$

Therefore $q \approx \frac{1}{45}$ and $2pq \approx \frac{1}{23}$ (4.35 per cent).

Thus about 1 in 23 individuals are heterozygous carriers.

(b) The equilibrium frequency for a recessive lethal allele is equal to the square root of the mutation rate; that is

$$q_e = \sqrt{\left(\frac{\mu}{S}\right)}$$

where $S = 1$.

Thus $\mu = 1/2\,000$ or 5×10^{-4} per generation.

This mutation rate is about ten times higher than a likely natural mutation rate.

(c) The appropriate model is

Genotype	A_1A_1	A_1A_2	A_2A_2
Fitness	$1 - S_1$	1	$1 - S_3$

If in equilibrium under heterozygous advantage, the frequency of A_2 is

$$q_e = \frac{S_1}{S_1 + S_3}$$

Now $q_e = 0.02$ and $S_3 = 1.0$. Therefore

$$0.02 = \frac{S_1}{S_1 + 1.0}$$

and

$$S_1 = \frac{0.02}{1 - 0.02} \approx 0.02$$

The relative fitnesses are therefore

Genotype	A_1A_1	A_1A_2	A_2A_2
Fitness	0.98	1.0	0

In other words, FCD can be maintained at its high frequency if the heterozygotes enjoy a selective advantage as small as about 2 per cent over the normal homozygotes.

(d) We again apply the equation

$$q_e = \frac{S_1}{S_1 + S_3}$$

where $S_1 = 1 - 0.9 = 0.1$, and $S_3 = 1.0$. Therefore

$$q_e = \frac{0.1}{1.1} = 0.090\,9$$

At this equilibrium allele frequency the incidence of affected births would be

$$q_e^2 = 0.008\,3$$

or about eight in every 1 000 births.

The segregation load is calculated from $S_1 p_e^2 + S_3 q_e^2$ (expression 20):

$$(0.1)(0.909\,1)^2 + (1.0)(0.008\,3)$$
$$0.082\,6 + 0.008\,3$$
$$= 0.090\,9$$

Although this means that the over-all fitness of the population is somewhat reduced, remember that the majority of this load is attributable to the large number of A_1A_1 individuals whose fitness is little impaired.

Note that even if there is a trend towards a higher equilibrium frequency of this sort— that is, about eight affected children in every 1 000 births—the approach is so slow that it would take at least a couple of thousand years.

Bibliography and references

General reading

For those of you who have found Units 9 and 10 particularly interesting, we suggest the following for further reading.

Ford, E. B. (4th edn 1975) *Ecological Genetics*, Chapman & Hall.

Mather, K. (1973) *Genetical Structure of Populations*, Chapman & Hall.

*Sheppard, P. M. (4th edn 1975) *Natural Selection and Heredity*, Hutchinson.

Williams, G. C. (1966) *Adaptation and Natural Selection*, Princeton University Press.

Quantitative aspects

Crow, J. F., and Kimura, M. (1970) *An Introduction to Population Genetics Theory*, Harper & Row.

*Wilson, E. O., and Bossert, V. H. (1971) *A Primer of Population Biology*, Sinauer,

Acknowledgements

Grateful acknowledgement is made to the following for material used in these Units:

Text

G. H. Hardy (1908), 'Mendelian proportions in a mixed population', *Science, N.Y.*, **28**; B. Wallace (1963), 'The elimination of an autosomal lethal from an experimental population of *D. melanogaster*', *American Naturalist*, **97**.

Figures

Figure 3 from A. E. Mourant *et al.* (2nd edn 1976), *The Distribution of the Human Blood Groups and Other Polymorphisms*, Oxford University Press; *Figure 5* and *Figure 1* based on B. Wallace *op cit.*; *Figure 7, 8 and 16* by courtesy of the British Museum of Natural History; *Figure 9* from H. B. D. Kettlewell (1965), 'Insect survival and selection for pattern', *Science, N.Y.*, **148**; *Figures 11, 12 and 14* from J. New (1958), 'A population study of *S. arvensis*', *Annals of Botany*, **22**; *Figure 13* redrawn from the *New Oxford Atlas*, 1975; *Figure 15* from J. van Zandt Brower, *American Naturalist* (1960), University of Chicago Press; *Figure 17* from Th. Dobzhansky and A. H. Sturtevant (1938), 'Inversions in the chromosomes of *D. pseudoobscura*', *Genetics*, **23**; *Figure 18* from Th. Dobzhansky (1948), 'Genetics of natural populations XVI', *Genetics*, **33**; *Figure 19* from L. E. Mettler and T. G. Gregg (1969), *Population Genetics and Evolution*, Prentice-Hall.

Tables

Table 4 from F. Vogel and M. R. Chakravartti (1966), *Humangenetik*, **3**; *Table 8 and 1* from B. Wallace *op cit.*; *Tables 9, 10 and 11* from H. B. D. Kettlewell (1955 and 1956), *Heredity*, **9** and **10**; *Tables 12 and 13* from Z. H. Abedi and A. W. A. Brown (1960), *Canadian Journal of Genetics and Cytology*, **2**; *Table 14* from N. H. Horowitz (1961), *Cold Spring Harbor Symposia in Quantitative Biology*, **26**; *Table 15* from J. van Zandt Brower *op cit.*; *Tables 16* from Th. Dobzhansky (1947), *Heredity*, **1**; *Table 17* from J. A. Beardmore Th. Dobzhansky and O. A. Pavlovsky (1960), *Heredity*, **14**; *Table 18* from Th. Dobzhansky and O. A. Pavlovsky (1962), *Heredity*, **16**.

* These books are written at the next level above that of the Units while the others are more advanced, particularly that by Crow and Kimura.

List of key symbols, expressions and equations

p The frequency of the A_1 allele.

q The frequency of the A_2 allele.

$P, 2Q, R$ The relative frequencies of the A_1A_1, A_1A_2 and A_2A_2 genotypes, respectively.

p_0, q_0 The frequencies of the A_1 and A_2 alleles before the first cycle of reproduction.

p_1, q_1 The frequencies of the A_1 and A_2 alleles immediately before the second cycle of reproduction.

p_n, q_n The frequencies of the A_1 and A_2 alleles after n generations.

p_e, q_e The frequencies of the A_1 and A_2 alleles at equilibrium. You will notice that in some texts $\hat{p}$ and $\hat{q}$ are used where we have used p_e and q_e: in statistics $\hat{p}$ and $\hat{q}$ are used to indicate *estimates* rather than equilibrium frequencies and so we have chosen to stick to p_e and q_e. Moreover, the e subscript is more descriptive than the circumflex.

(1) $p + q = 1$ When a gene is represented by two alleles the sum of the frequencies of these two alleles must be unity.

$(p + q)^2 = p^2 + 2pq + q^2$

If $p + q = 1$, then $p = 1 - q$

$p + 2q = p + q + q = 1 + q$

$p(q - p) = pq - p^2$

$\qquad = -p^2 + pq = -p(p - q)$

Just a few equations and rules you need to remember.

(2) $q = \sqrt{q^2}$ The frequency of a recessive allele is equal to the square root of the frequency of the recessive homozygote.

(3) $q_1 = \dfrac{q_0}{1 + q_0}$ The frequency of a recessive lethal allele in terms of its frequency in the previous generation.

(4) $q_n = \dfrac{q_0}{1 + nq_0}$ The frequency of a recessive lethal allele in the nth generation in terms of its frequency in the original generation.

(5) $n = \dfrac{1}{q_n} - \dfrac{1}{q_0}$ The number of generations required to change the frequency of a recessive lethal allele from a value q_0 to a value q_n.

(6) $\Delta q = -\dfrac{q^2}{1 + q}$ The change in frequency of a recessive lethal allele between generations in terms of its frequency in the earlier generation.

(7) $\bar{w} = 1 - S_1 p^2 - 2S_2 pq - S_3 q^2$ The mean fitness of a population as predicted from the generalized selection model.

(8) $\Delta q = \dfrac{pq}{\bar{w}} [p(S_1 - S_2) - q(S_3 - S_2)]$ The change in the frequency of an allele between generations allowing for selection against the three genotypes and for different allele frequencies.

(9) $q_e = \dfrac{S_2 - S_1}{2S_2 - S_1 - S_3}$ The equilibrium frequency of an allele on the basis of the generalized selection model.

(10) $\Delta q = -\dfrac{Spq^2}{1 - Sq^2}$ The change in allele frequency between generations for a recessive disadvantageous allele.

(11) $\Delta q = -\dfrac{Sp^2 q}{1 - S(2pq + q^2)}$ The change in allele frequency between generations for a dominant disadvantageous allele.

(12) $\Delta q = -\dfrac{Spq[h(p - q) + q]}{1 - 2Shpq - Sq^2}$ The change in allele frequency between generations for an allele having a fitness in the heterozygote intermediate between the two homozygotes.

(13) $\Delta q = \dfrac{pq(S_1 p - S_3 q)}{1 - S_1 p^2 - S_3 q^2}$ The change in allele frequency between generations for an allele showing heterozygous advantage for fitness.

(14) $\Delta q = -\dfrac{Spq^2(2q^2 - 1)}{1 - S(1 - q^2)^2 - Sq^4}$ The change in allele frequency between generations for frequency-dependent selection against a recessive disadvantageous allele.

$\bar{w} = 1 - S(p^4 + 4p^2 q^2 + q^4)$ The mean fitness of a population under frequency-dependent selection without dominance.

(16) $q = \sqrt{\left(\dfrac{\mu}{S}\right)}$ The frequency of a recessive disadvantageous allele maintained in a population by a balance between mutation and selection.

(17) $q_n - q_d = (1 - m)^n (q_0 - q_d)$ The difference in allele frequency between a recipient (immigrant) and a donor population after n generations equals $(1 - m)^n$, where m is the migration rate, times the initial difference in the frequency of that allele between the two populations.

(18) $\dfrac{q_n - q_d}{q_0 - q_d}$ The expression for estimating the proportion of genes in a migrant population that was derived from their ancestral population. $q_n - q_d$ is the difference in allele frequencies between the migrants and the donor population after n generations, and $q_0 - q_d$ is the difference between the allele frequencies of the original migrants (that is, the allele frequency in their ancestral population) and the donor population.

(19) $\dfrac{w_{max} - \bar{w}}{w_{max}}$ The formula for calculating the genetic load.

(20) $S_1 p^2 + S_3 q^2$ The segregation load for a pair of alleles alternative at a locus.